Principles of
Mathematical
Problem Solving

Principles of

Mathematical

Problem Solving

Martin J. Erickson and Joe Flowers

Truman State University

PRENTICE HALL
Upper Saddle River, New Jersey 07458

Library of Congress Cataloging-in-Publication Data

Erickson, Martin J.,
    Principles of mathematical problem solving / Martin J. Erickson
and Joe Flowers.
        p.   cm.
    Includes bibliographical references and index.
    ISBN 0-13-096445-X
    1. Problem solving.   I. Flowers, Joe.   II. Title.
    QA63.E75   1999   98-8331
    510–dc21   CIP

Cover art: *Portal Einer Moschee*, Paul Klee (1879–1940)

Editorial director, Tim Bozik
Editor-in-chief, Jerome Grant
Acquisition editor, George Lobell
Executive managing editor, Kathleen Schiaparelli
Managing editor, Linda Mihatov Behrens
Editorial assistants, Gale Epps, Nancy Bauer
Assistant VP production/manufacturing, David W. Riccardi
Manufacturing manager, Trudy Pisciotti
Manufacturing buyer, Alan Fischer
Creative director, Paula Maylahn
Art director, Jane Conte
Marketing manager, Melody Marcus

Printed in the United States of America

10 9 8 7 6 5 4 3 2

ISBN 0-13-096445-X

Prentice-Hall International (UK) Limited, London
Prentice-Hall of Australia Pty, Limited, Sydneypppp
Prentice-Hall Canada, Inc., Toronto
Prentice-Hall Hispanoamericana, S.A., New Delhi
Prentice-Hall of Japan, Inc., Tokyo
Prentice-Hall Pte.Ltd., Singapore
Editora Prentice-Hall do Brasil, Ltda., Rio de Janeiro

Contents

Preface

> As the strong man exults in his physical ability, delighting in such exercises as call his muscles into action, so glories the analyst in that moral activity which *disentangles*. He derives pleasure from even the most trivial occupations bringing his talent into play. He is fond of enigmas, of conundrums, hieroglyphics; exhibiting in his solutions of each a degree of *acumen* which appears to the ordinary apprehension præternatural. His results, brought about by the very soul and essence of method, have, in truth, the whole air of intuition.

EDGAR ALLEN POE
The Murders in the Rue Morgue, 1841

There is real joy in solving challenging mathematical problems. And while sometimes solutions may appear to be like magic, in most cases they are based on principles that can be learned and practiced. This book is about such principles. Each chapter introduces and explains a specific problem solving method (with examples), then presents a set of exercises and complete solutions, and finally poses an additional set of problems to challenge the reader. We believe that by studying the principles and applying them to the exercises, the reader will gain in problem solving ability and also in general mathematical insight. Eventually, he or she will be able to produce results that have "the whole air of intuition."

This book can serve as a text for problem solving courses or as a guide for individual study. It can be used by undergraduate students preparing for contests such as the William Lowell Putnam Mathematical Competition, and by advanced high school students studying for the American Mathematical Olympiad or regional competitions. It can also be read without any course or contest in mind—just for the pleasure of working on some interesting problems.

We trust that the problem sets include, for every reader, some old favorites and some new gems. Most of the problems have elegant, rather than tedious, solutions. Many of them also illustrate significant mathematical ideas. Therefore, in order to provide context, we have included a moderate amount of "theory" in some of the chapters. Solving problems is particularly rewarding when one discovers connections between the problems and other topics in mathematics.

The chapters of this book, and the exercises within each chapter, are arranged generally in order of increasing difficulty. Thus, the first two chapters, on generation of data and direct and indirect reasoning, should be easy going for most readers, and the second-to-last chapter, on the use of unforseen and surprising elements in a solution, should be quite challenging. The last chapter presents a list of problems that can be solved by using the methods described in the text.

Most chapters assume that the reader has studied material from previous chapters. For example, when solving problems using the pigeonhole principle (Chapter 9), it is sometimes necessary to use parity arguments (Chapter 7). Proof by contradiction is introduced early (in Chapter 3) and used often; the same is true of mathematical induction (introduced in Chapter 4). While many of the terms used in the text are defined in the glossary, additional background material on mathematical analysis, discrete structures, and abstract algebra can be found in Rudin's *Principles of Mathematical Analysis*, Johnsonbaugh's *Discrete Mathematics*, and Herstein's *Abstract Algebra*. Some excellent sources for more problems and problem solving techniques are listed in the bibliography.

We thank the following colleagues for their generous help in supplying problems and solutions: Robert Dobrow, Michael Adams, Dumont Hixson, Larry Wayne, D. W. Masser, Mark Bowron, Timothy Hamlin, Ross Honsberger, János Pach, Pankaj Agarwal, Murray Klamkin, Masaki Nishikawa, Sue Novinger, John Erickson, Thomas Hales, Ken Price, Jim Denvir, Suren Fernando, Rod Doll, Ke Tao, P. T. Bateman, Evan Haffner, Joe Hemmeter, Frank Sottile, Mark Schlatter, Olga Yiparaki, Hugh Montgomery, Mel Hochster, Todd Hammond, George Ashline, Mansur Boase, H. Chad Lane, Dan Jordan, Gregory C. Jones, Don Bindner, Loukas Grafakos, Eric Norige, Aaron Nord, Rebecca Sharpe, Weiwei Zhu, and David Hardy.

M. E. and J. F.

Chapter 1

Data

When solving problems it is often best to begin by obtaining data. Gathering data can be as important in a mathematical investigation as in a scientific investigation, and sometimes just as necessary in getting started.

For example, when working on a problem about a sequence of integers, try listing some elements of the sequence. Do the numbers follow a pattern? Are they related to another, more familiar sequence?

Consider the following question about lines in the plane.

Example 1.1. Suppose that there are n lines in the plane, no two parallel and no three intersecting at a point. Into how many regions is the plane divided by these lines?

Solution: First we generate some data. Let $f(n)$ be the number of regions into which the plane is divided by n lines. With some simple sketches (Figure 1.1), we see that $f(1) = 2$, $f(2) = 4$, $f(3) = 7$, and $f(4) = 11$. What can we say about the progression of numbers 2, 4, 7, 11, ...? The successive differences of these numbers are 2, 3, 4, ... Thus it appears that $f(n) - f(n-1) = n$, or, equivalently, $f(n) = f(n-1) + n$, for $n \geq 2$.

Conjecture: $f(n) = f(n-1) + n$, for $n \geq 2$.

Is our conjecture true, and why? Notice in Figure 1.1 that when we have $n-1$ lines and add an nth line, the nth line must intersect all the

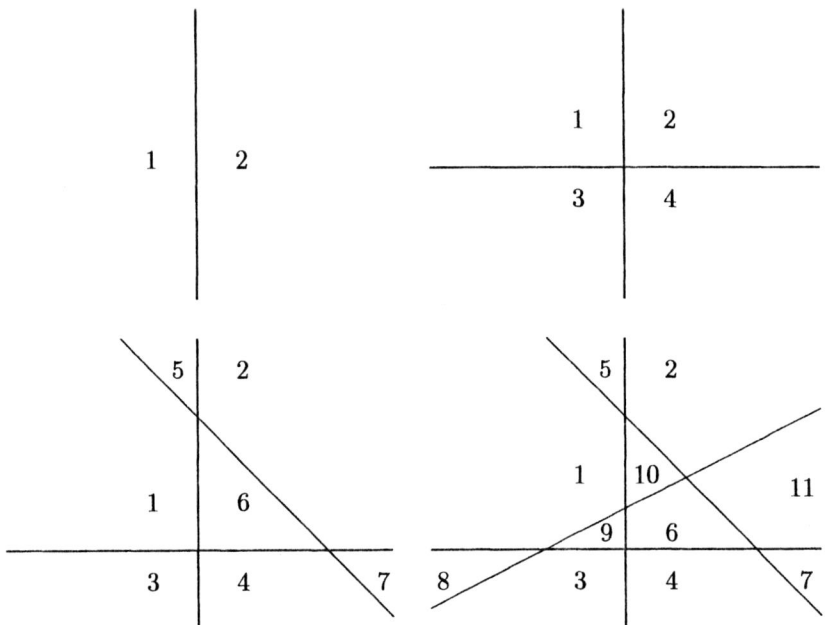

Figure 1.1: One, two, three, and four lines dividing the plane into two, four, seven, and eleven regions.

others. Therefore the nth line is divided into n segments by the other lines. Since each of these segments divides an old region into two new regions, the addition of the nth line increases the number of regions by n. This explains the rate of growth of $f(n)$; we have proved our conjecture that $f(n) = f(n-1) + n$.

The formula $f(n) = f(n-1) + n$ is a *recurrence relation* for $f(n)$. That is, we can use it to calculate $f(n)$ for successively greater values of n. Although this is good, we might also want an explicit formula for $f(n)$, i.e., one that allows us to compute $f(n)$ immediately for any given value of n. How do we obtain such a formula? Due to its particularly simple form, our recurrence relation can be solved by "working backwards":

$$
\begin{aligned}
f(n) &= f(n-1) + n \\
     &= f(n-2) + (n-1) + n \\
     &= f(n-3) + (n-2) + (n-1) + n
\end{aligned}
$$

$$\vdots$$

$$= \quad f(1) + (2 + 3 + 4 + \cdots + n)$$

$$= \quad 2 + \frac{n(n+1)}{2} - 1$$

$$= \quad \frac{n^2 + n + 2}{2}.$$

We now have an explicit formula:

$$f(n) = \frac{n^2 + n + 2}{2}.$$

Finally, it is a good idea to "double-check" an answer to see that it fits the data. So we put the numbers 1, 2, 3, 4 into our formula for $f(n)$ and obtain the values 2, 4, 7, 11, as expected. And now that we have a formula for $f(n)$, we can use it to obtain values that we didn't know previously. Into how many regions is the plane divided by 100 lines in general position? Just use the formula.

Note: We can also obtain the formula for $f(n)$ from Euler's relation,

$$V + F = E + 2,$$

for a connected planar graph with V vertices, F faces, and E edges (see Glossary). We turn the given configuration of lines into a graph by placing a vertex at each line intersection and joining all the unbounded lines at a single new vertex outside the convex hull of the other vertices. (We must bend lines to do this but that is allowed in a graph.) Now we apply Euler's formula to determine $F = f(n)$. Since $V = \binom{n}{2} + 1$ and $E = n^2$, we find that

$$F = n^2 + 2 - \frac{n(n-1)}{2} - 1 = \frac{n^2 + n + 2}{2}.$$

■

Example 1.2. (Putnam Competition, 1990; modified) Let

$$T_0 = 2, \ T_1 = 3, \ T_2 = 6,$$

and for $n \geq 3$,

$$T_n = (n+4)T_{n-1} - 4nT_{n-2} + (4n-8)T_{n-3}.$$

The first few terms are

$$2, \ 3, \ 6, \ 14, \ 40, \ 152, \ 784, \ 5168, \ 40576, \ 363392.$$

Find, with proof, a formula for T_n of the form $T_n = A_n + B_n$, where $\{A_n\}$ and $\{B_n\}$ are well-known sequences.

Solution: Although we are given the first terms of the sequence $\{T_n\}$, there is more investigative work to do, unless we want to try to solve the recurrence relation directly (it's possible but difficult). What "well-known" sequences do you know?

Some sequences that come to mind are

$$
\begin{aligned}
\{n\} &= \{0, 1, 2, 3, 4, 5, 6, 7, \ldots\} \\
\{2n\} &= \{0, 2, 4, 6, 8, 10, 12, 14, \ldots\} \\
\{2n + 1\} &= \{1, 3, 5, 7, 9, 11, 13, 15, \ldots\} \\
\{n^2\} &= \{0, 1, 4, 9, 16, 25, 36, 49, \ldots, \} \\
\{(-1)^n\} &= \{1, -1, 1, -1, 1, -1, 1, -1, \ldots\} \\
\{2^n\} &= \{1, 2, 4, 8, 16, 32, 64, 128, \ldots\} \\
\{n!\} &= \{1, 1, 2, 6, 24, 120, 720, 5040, \ldots\}.
\end{aligned}
$$

It doesn't take long to see that we can let $A_n = 2^n$ and $B_n = n!$. All that was required was the generation of some data.

Conjecture: $T_n = 2^n + n!$, for all $n \geq 0$.

We prove the conjecture using mathematical induction, having already observed that the formula holds for $n = 0$, 1, and 2. (The reader may wish to skip the proof until after looking at Chapter 4.) Assume that the formula holds for $n \leq k$. Then

$$
\begin{aligned}
T_{k+1} &= (k+5)T_k - 4(k+1)T_{k-1} + (4k-4)T_{k-2} \\
&= (k+5)(2^k + k!) - 4(k+1)(2^{k-1} + (k-1)!) \\
&\quad + 4(k-1)(2^{k-2} + (k-2)!) \\
&= 2^{k-2}(4k + 20 - 8k - 8 + 4k - 4) \\
&\quad + (k-2)![(k+5)(k-1)k - 4(k+1)(k-1) + 4(k-1)] \\
&= 2^{k+1} + (k+1)!.
\end{aligned}
$$

Hence the formula holds for $n \geq 0$. ∎

Example 1.3. How many ways can a $1 \times n$ rectangle be filled by nonoverlapping 1×1 and 1×2 rectangles?

1 × n rectangle   1 × 1 rectangle   1 × 2 rectangle

Solution: Let $g(n)$ be the number of such fillings. We generate some data.

$g(1) = 1$

$g(2) = 2$

$g(3) = 3$

$g(4) = 5$

$g(5) = 8$

Thus

$$\{g(n)\} = \{1, 2, 3, 5, 8, \ldots\}.$$

The Fibonacci numbers are defined by

$$f_0 = 0, \quad f_1 = 1, \quad \text{and}$$
$$f_n = f_{n-1} + f_{n-2}, \quad \text{for } n \geq 2.$$

Thus

$$\{f_n\} = \{0, 1, 1, 2, 3, 5, 8, \ldots\}.$$

It seems reasonable to conjecture that $g(n)$ is equal to the $(n + 1)$st Fibonacci number, f_{n+1}.

Conjecture: $g(n) = f_{n+1}$.

We test our conjecture by determining $g(6)$ and find that, indeed, $g(6) = f_7 = 13$.

$g(6) = 13$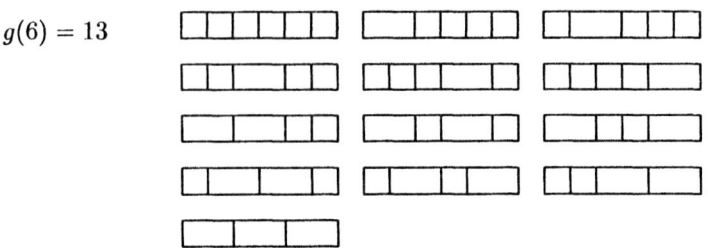

We have already seen that the sequence $\{g(n)\}$ agrees with the Fibonacci number f_{n+1} when $n = 1$ and $n = 2$: $g(1) = f_2 = 1$ and $g(2) = f_3 = 2$. If we can show that $\{g(n)\}$ satisfies the same recurrence relation as does the Fibonacci sequence, then it will follow that the two sequences are identical. In calculating $g(n)$, there are two cases: the rightmost unit in the filling of the $1 \times n$ rectangle is a 1×1 rectangle or else it is a 1×2 rectangle. In the first case, the remainder of the rectangle can be filled in $g(n-1)$ ways. In the second case, it can be filled in $g(n-2)$ ways. Therefore, $g(n) = g(n-1) + g(n-2)$. Hence $g(n) = f_{n+1}$. ∎

In solving the following problems, the goal is to generate data and discover patterns. We are more concerned with making correct conjectures than with proving them.

Problems

1.1 Let $f(x) = xe^{2x}$. Let $f^{(n)}$ be the nth derivative of f. Show that $f^{(n)} = a_n e^{2x} + b_n x e^{2x}$, for some numbers a_n and b_n. Find formulas for a_n and b_n.

1.2 Suppose that $n \geq 3$. Show that an even number of the fractions

$$\frac{1}{n}, \frac{2}{n}, \frac{3}{n}, \ldots, \frac{n-1}{n}$$

are in lowest terms.

1.3 Find a formula for

$$\prod_{i=1}^{n} \left\{ 1 - \frac{4}{(2i-1)^2} \right\}.$$

1.4 Find a formula for

$$f_1 - f_2 + f_3 - f_4 + f_5 - f_6 + \cdots + (-1)^{n+1} f_n,$$

where f_n is the nth Fibonacci number.

1.5 In a certain jail, there are 100 cells, numbered 1 through 100, all locked. During the night, the jailer turns the key on every cell, thereby unlocking them. Then he turns the key on every second cell (2, 4, 6, ...), locking them. Then he turns the key on every third cell (3, 6, 9, ...), and so on, so that on his last pass by the cells, he turns the key on only the 100th cell. Which cells are now open?

1.6 Let $b(n)$ be the number of binary strings of length n which do not contain the substring 11. Find a formula for $b(n)$. For example, $b(3) = 5$ (the five strings are 000, 001, 010, 100, and 101).

1.7 A *composition* of a positive integer n is a summation $n = n_1 + \cdots + n_k$, for some k, where the summands n_i are positive integers and the order of the summands matters. Let $S(n)$ be the number of compositions of n. Find a formula for $S(n)$.

1.8 (From Lewis Carroll.) Find three right triangles with sides of integer length all having the same area.

Solutions

1.1 We generate some data:

$$
\begin{aligned}
f^{(1)}(x) &= e^{2x} + 2xe^{2x} \\
f^{(2)}(x) &= 4e^{2x} + 4xe^{2x} \\
f^{(3)}(x) &= 12e^{2x} + 8xe^{2x} \\
f^{(4)}(x) &= 32e^{2x} + 16xe^{2x} \\
f^{(5)}(x) &= 80e^{2x} + 32xe^{2x}.
\end{aligned}
$$

It appears that
$$ f^{(n)} = a_n e^{2x} + b_n x e^{2x}, $$
where $a_n = n2^{n-1}$ and $b_n = 2^n$. To prove that these formulas are correct, we use induction. Note that $a_1 = 1 = 1 \cdot 2^{1-1}$ and $b_1 = 2 = 2^1$. Now assume that $a_n = n2^{n-1}$ and $b_n = 2^n$. Then
$$ f^{(n)}(x) = n2^{n-1}e^{2x} + 2^n x e^{2x} $$

and

$$
\begin{aligned}
f^{(n+1)}(x) &= n2^{n-1} \cdot 2e^{2x} + 2^n e^{2x} + 2^n x \cdot 2e^{2x} \\
&= (n+1)2^n e^{2x} + 2^{n+1} x e^{2x}.
\end{aligned}
$$

Hence $a_{n+1} = (n+1)2^n$ and $b_{n+1} = 2^{n+1}$, which proves our claim.

1.2 We generate some data. For $3 \leq n \leq 8$, we list the fractions k/n, with $1 \leq k \leq n-1$ and k and n are relatively prime:

$$n = 3 \qquad \tfrac{1}{3}, \ \tfrac{2}{3}$$

$$n = 4 \qquad \tfrac{1}{4}, \ \tfrac{3}{4}$$

$$n = 5 \qquad \tfrac{1}{5}, \ \tfrac{2}{5}, \ \tfrac{3}{5}, \ \tfrac{4}{5}$$

$$n = 6 \qquad \tfrac{1}{6}, \ \tfrac{5}{6}$$

$$n = 7 \qquad \tfrac{1}{7}, \ \tfrac{2}{7}, \ \tfrac{3}{7}, \ \tfrac{4}{7}, \ \tfrac{5}{7}, \ \tfrac{6}{7}$$

$$n = 8 \qquad \tfrac{1}{8}, \ \tfrac{3}{8}, \ \tfrac{5}{8}, \ \tfrac{7}{8}.$$

For each n, the list contains an even number of fractions. In fact, we see that the fractions occur in pairs. If k/n is in the list, then so is $(n-k)/n$. To prove this fact is not difficult. If $d|k$ and $d|n$, then $d|(n-k)$. And if $d|(n-k)$ and $d|n$, then $d|k$. (The notation $d|k$ means that d divides k evenly.) Hence k and n are relatively prime precisely when $n-k$ and n are. Therefore, an even number of fractions k/n are in lowest terms. (Note that we have $k/n = (n-k)/n$ only when $n = 2k$, and the fraction $k/(2k)$ is definitely not in lowest terms.)

1.3 Let x_n be the quantity in question. We compute the first few values of x_n: $x_1 = -3/1$, $x_2 = -5/3$, $x_3 = -7/5$. These data suggest the formula $x_n = -(2n+1)/(2n-1)$, and this formula is proved in the solution to Problem 4.4 (in the chapter on induction).

1.4 Let
$$g_n = f_1 - f_2 + f_3 - f_4 + f_5 - f_6 + \cdots + (-1)^{n+1} f_n.$$

Then $g_1 = 1$, $g_2 = 0$, $g_3 = 2$, $g_4 = -1$, $g_5 = 4$, $g_6 = -4$, $g_7 = 9$, and $g_8 = -12$. We see that g_n alternates in sign (starting with $n = 3$). Furthermore, the absolute value of g_n always differs by 1 from the Fibonacci number f_{n-1}. Putting these observations together, we guess that
$$g_n = (-1)^{n+1} f_{n-1} + 1$$

for $n \geq 1$. The conjecture is easily proved by induction. For $n = 1$ we have $g_1 = 1 = (-1)^2 f_0 + 1$. Assume that the formula holds for n. Then

$$\begin{aligned} g_{n+1} &= g_n + (-1)^{n+2} f_{n+1} \\ &= (-1)^{n+1} f_{n-1} + 1 + (-1)^{n+2} f_{n+1} \end{aligned}$$

$$
\begin{aligned}
&= (-1)^{n+2} \left(f_{n+1} - f_{n-1} \right) + 1 \\
&= (-1)^{n+2} f_n + 1.
\end{aligned}
$$

We conclude that the formula holds for all $n \geq 1$.

1.5 We check some values of n to see whether cell n ends up open or closed.

$n = 1$	open
$n = 2$	closed
$n = 3$	closed
$n = 4$	open
$n = 5$	closed
$n = 6$	closed
$n = 7$	closed
$n = 8$	closed
$n = 9$	open
$n = 10$	closed

From these data it appears that cell n is open if and only if n is a perfect square. To test this hypothesis we check a few more values of n.

$n = 11$	closed
$n = 12$	closed
$n = 13$	closed
$n = 14$	closed
$n = 15$	closed
$n = 16$	open
$n = 17$	closed
$n = 18$	closed
$n = 19$	closed
$n = 20$	closed

Now we feel confident that cell n is open if and only if n is a perfect square. This fact is proved in Example 7.4 (in the chapter on parity).

1.6 We generate some data and find that

$$
\{b(n)\} = \{2, 3, 5, 8, 13, \ldots\}.
$$

Thus we are led to the conjecture that $b(n) = f_{n+2}$. This is easy to prove via the result of Example 1.3, i.e., that f_{n+1} is the number of fillings of an $n \times 1$ rectangle with 1×1 and 2×1 rectangles. Such fillings correspond to binary strings of length $n+1$ using substrings 10 and 0. Discarding the rightmost bit (necessarily a 0) of each of these strings, we obtain the desired strings of length n. The correspondence is clearly reversible.

1.7 We generate some data.

$$
\begin{aligned}
S(1) &= 1 & &1 \\
S(2) &= 2 & &2,\ 1+1 \\
S(3) &= 4 & &3,\ 1+2,\ 2+1,\ 1+1+1 \\
S(4) &= 8 & &4,\ 1+3,\ 3+1,\ 2+2,\ 1+1+2, \\
& & &1+2+1,\ 2+1+1,\ 1+1+1+1 \\
S(5) &= 16 & &5,\ 1+4,\ 4+1,\ 2+3,\ 3+2,\ 1+1+3, \\
& & &1+3+1,\ 3+1+1,\ 1+2+2,\ 2+1+2, \\
& & &2+2+1,\ 1+1+1+2,\ 1+1+2+1, \\
& & &1+2+1+1,\ 2+1+1+1,\ 1+1+1+1+1
\end{aligned}
$$

It appears that $S(n) = 2^{n-1}$. To prove it, we note that each composition of n can be obtained from the composition consisting of n 1's,

$$ n = \underbrace{1 + 1 + \cdots + 1}_{n}. $$

Between each consecutive pair of 1's we have the option of placing a vertical line, and, summing the elements bounded by lines, we obtain a composition. For example,

$$ 9 = 1 + 1 + 1 \mid +1 \mid +1 + 1 + 1 \mid +1 + 1 $$

gives the composition $9 = 3 + 1 + 3 + 2$. As there are, in general, $n - 1$ places for the vertical lines, there are 2^{n-1} compositions.

Note: Here is another proof. A composition of n into k parts may be regarded as a distribution of n indistinguishable objects into k nonempty classes. Without the requirement that the classes be nonempty, the number of distributions is $\binom{n+k-1}{k-1}$. Requiring the classes to be nonempty effectively reduces the number of objects to be distributed by k. Therefore, the number of distributions of n objects into k nonempty classes is $\binom{n-1}{k-1}$. It follows that

$$ S(n) = \sum_{k=1}^{n} \binom{n-1}{k-1} = 2^{n-1}. $$

1.8 Triples (a, b, c) of positive integers satisfying $a^2 + b^2 = c^2$ are called *Pythagorean triples*. The corresponding triangle is called a *Pythagorean triangle*. Noting the identity

$$ (u^2 - v^2)^2 + 4u^2v^2 = (u^2 + v^2)^2, $$

we can use the formulas $a = u^2 - v^2$, $b = 2uv$, $c = u^2 + v^2$ to generate some Pythagorean triples:

u	v	a	b	c	area
2	1	3	4	5	6
3	1	8	6	10	24
3	2	5	12	13	30
4	1	15	8	17	60
4	2	12	16	20	96
4	3	7	24	25	84	etc.

Continuing the table, we soon generate the following data:

u	v	a	b	c	area
7	3	40	42	58	840
7	5	24	70	74	840
8	7	15	112	113	840

Note: It can be shown that all Pythagorean triples (a, b, c) (excluding interchanges of a and b) are generated without repetition by the formulas $a = k(u^2 - v^2)$, $b = k(2uv)$, $c = k(u^2 + v^2)$, where u, v, and k are positive integers, $u > v$, $\gcd(u, v) = 1$, and u and v are not both even or both odd.

Additional Problems

1.9 How many ways can a $2 \times n$ rectangle be filled by n 2×1 rectangles?

1.10 What is the maximum number of regions into which $\mathbf{R}^3$ can be divided by n planes?

1.11 What is the maximum number of regions into which the plane can be divided by n circles?

Note: Be careful! The data in this problem and the next one may be misleading.

1.12 What is the maximum number of regions into which $\mathbf{R}^3$ can be divided by n spheres?

1.13 Let $s(n)$ denote the number of ways of writing a positive integer n as the sum of odd positive integers, where the order of the summands matters. Find a formula for $s(n)$.

1.14 Find a formula for

$$f(n) = 1^2 + 3^2 + \cdots + (2n-1)^2.$$

1.15 Find a formula for

$$g(n) = 1^3 - 3^3 + 5^3 - 7^3 + \cdots + (-1)^{n-1}(2n-1)^3.$$

Hint: Observe that $n \mid g(n)$ for all $n \geq 1$.

1.16 Which prime numbers are sums of two square numbers?

1.17 (a) Which positive integers are sums of three squares?

(b) Which positive integers are sums of four squares?

Note: See Example 8.3.

1.18 (a) In the game Even Steven, two players, Alpha and Beta, divide a
pile of n sticks (n an odd number) into two piles—Alpha's pile and
Beta's pile. The division occurs by Alpha and Beta alternately taking
one or two sticks from the original pile. The object is to have, at the
end, an even number of sticks in one's own pile. Someone must win,
because whenever an odd number is the sum of two numbers, one of
the numbers is even and the other odd. Who wins Even Steven, given
ideal play by both players?

(b) The game Oddball is identical to Even Steven except that the
winner is the player who at the end has an odd number of sticks.
Who wins Oddball?

Chapter 2

Direct and Indirect Reasoning

"I think we're apt to overlook the simple explanations, which are, after all, nearly always the true ones."

BURTON E. STEVENSON
Mr. Godfrey, *The Holladay Case*, 1903

"The solving of almost every crime mystery depends on something which seems, at the first glance, to bear *no relation whatsoever* to the original crime."

ELSA BARKER
Dexter Drake, *The C. I. D. of Dexter Drake*, 1929

In this chapter we discuss two types of attack on problems. For some problems, a straightforward method is feasible. We do the most obvious thing and it works. This is the direct approach. For other problems, we must do some preliminary analysis before we understand what will constitute a complete solution. This is the indirect approach. We just start somewhere (using intuition or an educated guess, perhaps) and see where we are led.

Here are two simple examples of the direct approach.

Example 2.1. Prove that $n^4 + 4$ is a composite number for each integer $n \geq 2$.

Solution: We start by testing the claim for small values of n. For $n = 2$, 3, 4, 5, and 6, $n^4 + 4 = 20$, 85, 260, 629, and 1300, respectively. Since these numbers are composite ($20 = 4 \cdot 5$, $85 = 5 \cdot 17$, $260 = 10 \cdot 26$, $629 = 17 \cdot 37$, and $1300 = 13 \cdot 100$), we have confirmation of the claim for small values of n. But how do we prove the claim in general? Since a composite number is a number with nontrivial factors, we might ask whether the algebraic expression $n^4 + 4$ has nontrivial factors. If so, then the numbers given by the expression must be composite. So the problem amounts to trying to find a factorization of $n^4 + 4$. We observe that $n^4 + 4$ is the difference of two squares, and such a difference always factors. Thus we obtain

$$n^4 + 4 = (n^2 + 2)^2 - 4n^2 = (n^2 + 2n + 2)(n^2 - 2n + 2).$$

Since both of the above factors are greater than 1 for $n \geq 2$, $n^4 + 4$ is composite for these values of n.   ∎

Example 2.2. Prove that $|z| = 2|z - 3|$ if and only if $|z - 4| = 2$, where $z \in \mathbf{C}$.

Solution: Representing z as $x + yi$, we obtain two equations in x and y,

$$\sqrt{x^2 + y^2} = 2\sqrt{(x - 3)^2 + y^2}$$

and

$$\sqrt{(x - 4)^2 + y^2} = 2.$$

We must show that these equations are equivalent. But that practically happens by itself:

$$
\begin{aligned}
\sqrt{x^2 + y^2} &= 2\sqrt{(x - 3)^2 + y^2} \\
\Longleftrightarrow \quad x^2 + y^2 &= 4(x^2 - 6x + 9 + y^2) \\
\Longleftrightarrow \quad 0 &= 3x^2 - 24x + 3y^2 + 36 \\
\Longleftrightarrow \quad 4 &= (x - 4)^2 + y^2 \\
\Longleftrightarrow \quad 2 &= \sqrt{(x - 4)^2 + y^2}.
\end{aligned}
$$

∎

Some solutions are straightforward but at the same time quite difficult to find. The following problem provides a good illustration.

Example 2.3. Prove or disprove that

$$\sqrt{11 + 2\sqrt{29}} + \sqrt{16 - 2\sqrt{29} + 2\sqrt{55 - 10\sqrt{29}}} = \sqrt{5} + \sqrt{22 + 2\sqrt{5}}.$$

Solution: We prove the identity by showing that the left-hand side simplifies to the right-hand side. We first simplify the second term by observing that the quantity under the outer radical is actually a perfect square trinomial. Hence,

$$\sqrt{11 + 2\sqrt{29}} + \sqrt{16 - 2\sqrt{29} + 2\sqrt{55 - 10\sqrt{29}}}$$

$$= \sqrt{11 + 2\sqrt{29}} + \sqrt{\left(11 - 2\sqrt{29}\right) + 2\sqrt{5}\sqrt{11 - 2\sqrt{29}} + 5}$$

$$= \sqrt{11 + 2\sqrt{29}} + \sqrt{\left(\sqrt{11 - 2\sqrt{29}} + \sqrt{5}\right)^2}$$

$$= \sqrt{11 + 2\sqrt{29}} + \sqrt{11 - 2\sqrt{29}} + \sqrt{5}$$

$$= \sqrt{5} + \sqrt{\left(\sqrt{11 + 2\sqrt{29}} + \sqrt{11 - 2\sqrt{29}}\right)^2}$$

$$= \sqrt{5} + \sqrt{11 + 2\sqrt{29} + 11 - 2\sqrt{29} + 2\sqrt{5}}$$

$$= \sqrt{5} + \sqrt{22 + 2\sqrt{5}}.$$

∎

In the following two problems, the method is indirect. Auxillary relations or results must be considered before the posed problems are dealt with.

Example 2.4. Show how to put ten digits in the boxes below to make a ten-digit number N with the property that the digit in the ith box is the number of occurrences of i in N.

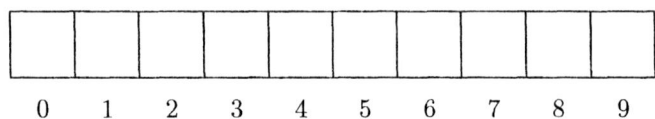

   0   1   2   3   4   5   6   7   8   9

Solution: Let the digits be $a_0, a_1, \ldots, a_9$. Knowing that a_i is the number of i's among the ten digits gives us two simple equations:

$$a_0 + a_1 + \cdots + a_9 = 10$$

$$0 \cdot a_0 + 1 \cdot a_1 + \cdots + 9 \cdot a_9 = 10.$$

Subtracting the second equation from the first yields

$$a_0 \quad = \quad a_2 + 2a_3 + 3a_4 + \cdots + 8a_9. \tag{2.1}$$

From the definition of a_0 we know that $a_0 \neq 0$. Also, $a_0 \neq 1$, for if $a_0 = 1$ then the equation (2.1) forces a_3, a_4, ..., a_9 to each be zero, which would imply $a_0 \geq 7$ (by definition). By similar reasoning, $a_0 \neq 2$. Suppose that $a_0 = m \geq 3$. Then, by definition, $a_m \geq 1$. Now (2.1) forces $a_m = 1$ and $a_n = 0$ for $n = 3$, ..., 9, $n \neq m$, and $a_2 = 1$. It follows that $a_1 = 2$ and there are six zeros in N, so that $m = 6$. We have found all of the a_i and they make the unique ten-digit number $N = 6210001000$.

6	2	1	0	0	0	1	0	0	0
0	1	2	3	4	5	6	7	8	9

■

Example 2.5. Do there exist odd integers d_1, d_2, ..., d_m with $1 < d_1 < d_2 < \cdots < d_m$ and

$$\sum_{i=1}^{m} \frac{1}{d_i} = 1?$$

Solution: If the answer were no, then there could be no odd perfect number, since for an odd perfect number n we would have

$$\sum_{\substack{d \mid n \\ d \neq 1}} \frac{1}{d} = 1.$$

In fact, no one knows whether or not there exists an odd perfect number.

But the answer is yes and we will produce such a set of integers. We start with the odd abundant number $945 = 3^3 \cdot 5 \cdot 7$. (Recall that n is *abundant* if the sum of the divisors of n is greater than $2n$.) To check that 945 is an abundant number, we note that the sum of its divisors is $\sigma(945) = 40 \cdot 6 \cdot 8 = 1920$, which is greater than $2 \cdot 945 = 1890$. Now we try to write 945 as a sum of some of its divisors. Noting that

$$\sum_{\substack{d \mid 945 \\ d \neq 945}} d = 1920 - 945 = 975 = 945 + 30,$$

and 9 and 21 are divisors of 945 which sum to 30, it follows that

$$\sum_{\substack{d|945 \\ d \neq 9,21,945}} d = 945.$$

Hence

$$\sum_{\substack{d|945 \\ d \neq 9,21,945}} \frac{d}{945} = 1,$$

and this relation may be written

$$\sum_{\substack{e|945 \\ e \neq 1,45,105}} \frac{1}{e} = 1.$$

Thus the divisors of 945, excluding 1, 45, and 105, are a solution to the problem.

Question: It happens that 945 is the smallest odd abundant number. Does the type of construction above work starting with *every* odd abundant number? ■

Problems

2.1 Show that the equation $y^2 = x^4 + 1$ has only finitely many solutions in integers x, y, and find them all.

2.2 Do the same as above with the equation $y^2 = x^4 + x + 7$.

2.3 You have seen two of the Smith children countless times. Half of the time they both had blue eyes and half of the time they did not. What is the smallest possible number of Smith children and what color eyes do they have?

2.4 How many zeros occur at the end of the expression

$$\prod_{n=2}^{1000} (n^2 - 1)?$$

2.5 (a) Show that the equation

$$1 + x + \frac{x^2}{2!} + \cdots + \frac{x^{2n}}{(2n)!} = 0$$

has no real roots.

(b) Show that the equation

$$1 + x + \frac{x^2}{2!} + \cdots + \frac{x^{2n+1}}{(2n+1)!} = 0$$

has exactly one real root.

2.6 Prove that the quadrilateral $ABCD$ has perpendicular diagonals if and only if

$$AB^2 + CD^2 = AD^2 + BC^2.$$

2.7 Express the function $h(x) = 2x$ as a composition of functions $g(x) = 1/x$ and $f_t(x) = x + t$ for suitable real numbers t.

2.8 (Putnam Competition, 1964) Let S be a set of $n > 0$ elements, and let A_1, A_2, ..., A_k be a family of distinct subsets, with the property that any two of these subsets meet. Assume that no other subset of S meets all of the A_i. Prove that $k = 2^{n-1}$.

2.9 (Putnam Competition, 1996) Suppose that each of twenty students has made a choice of anywhere from zero to six courses from a total of six courses offered. Prove or disprove: There are five students and two courses such that all five have chosen both courses or all five have chosen neither.

2.10 Show that there exists an irrational number x such that $x^{\sqrt{2}}$ is rational.

2.11 Suppose that the polynomial $f(x) = ax^2 + bx + c$ satisfies $|f(x)| \le 1$ for all $|x| \le 1$. Prove that $|f'(x)| \le 4$ for all $|x| \le 1$.

2.12 Suppose that v_1, ..., v_n are points in $\mathbf{R}^d$ with the property that $|v_i - v_j| = 1$ for each distinct pair v_i, v_j. Show that $n \le d + 1$.

Solutions

2.1 Solution (1): Since $(y - x^2)(y + x^2) = 1$, either $y - x^2 = 1$ and $y + x^2 = 1$, or $y - x^2 = -1$ and $y + x^2 = -1$. In the first case, $x = 0$ and $y = 1$. In the second case, $x = 0$ and $y = -1$. Hence the only solutions are $(0, \pm 1)$.

Solution (2): If $x = 0$, then $y = \pm 1$. If $x \ne 0$, then

$$(x^2)^2 < x^4 + 1 < x^4 + 2x^2 + 1 = (x^2 + 1)^2,$$

so $y^2 = x^4 + 1$ lies between two consecutive squares, which is impossible. Again we find that the only solutions are $(0, \pm 1)$.

2.2 We use the idea of Solution (2) of the previous problem. If $x = 1$, then $y = \pm 3$. If $x = 2$, then $y = \pm 5$. If $x \geq 3$, then

$$(x^2)^2 < x^4 + x + 7 < x^4 + 2x^2 + 1 = (x^2 + 1)^2,$$

and, as in the solution above, this is impossible. It is easy to check that there are no solutions with $-7 \leq x \leq 0$. If $x \leq -8$, then

$$(x^2 - 1)^2 = x^4 - 2x^2 + 1 < x^4 + x + 7 < (x^2)^2,$$

and there are no solutions. Hence the only solutions are $(1, \pm 3)$ and $(2, \pm 5)$.

2.3 The minimum possible number of Smith children is four, and their eyes are blue, blue, blue, and green (green represents any non-blue color). With such a set of children, there are three pairs with blue eyes and three pairs who do not both have blue eyes. Hence the probability that both children in a given pair have blue eyes is $1/2$.

It is easy to see that fewer than four children do not suffice. With two children, the probability that they both have blue eyes is either 0 or 1. With three children, the probablility is 0 (if none or just one child has blue eyes), $1/3$ (if two children have blue eyes and one does not), or 1 (if all three have blue eyes).

2.4 Clearly, we want to determine the largest m for which

$$\prod_{n=2}^{1000} (n^2 - 1) \equiv 0 \bmod 5^m.$$

We have $n^2 - 1 = (n-1)(n+1) \equiv 0 \bmod 5^l$ if and only if $n = k5^l \pm 1$. In the range $n = 2, \ldots, 1000$, there are

$$\left\lfloor \frac{10^3}{5} \right\rfloor + \left\lfloor \frac{10^3}{25} \right\rfloor + \left\lfloor \frac{10^3}{125} \right\rfloor + \left\lfloor \frac{10^3}{625} \right\rfloor$$

contributions to m from n of the form $k5^l + 1$ and almost an equal contribution from n of the form $k5^l - 1$, except that in the latter case the factor 1000 is not attained for any $n - 1$. Hence the largest possible value of m is $200 + 40 + 8 + 1 + 199 + 39 + 7 + 1 = 495$.

2.5 (a) Let

$$f(x) = 1 + x + \frac{x^2}{2!} + \cdots + \frac{x^{2n}}{(2n)!}.$$

We will show that the local minima of f are positive. Since f is a polynomial of even degree, it will follow that f is positive for all real x. The minima of f can only occur when $f'(x) = 0$. However,

$$f'(x) = 1 + x + \frac{x^2}{2!} + \cdots + \frac{x^{2n-1}}{(2n-1)!},$$

so that

$$f(x) = f'(x) + \frac{x^{2n}}{(2n)!}.$$

Hence $f(x)$ is positive whenever $f'(x) = 0$. (Note that $f'(0) \neq 0$.) Therefore f has no real roots.

(b) Let

$$g(x) = 1 + x + \frac{x^2}{2!} + \cdots + \frac{x^{2n+1}}{(2n+1)!}.$$

Since $g(0) = 1$ and g is negative for sufficiently small x, it follows by the intermediate value theorem that g has a real root. Since $g'(x) = f(x) > 0$ for all real x, g is an increasing function. Hence, by Rolle's theorem, g does not have two (or more) real roots.

2.6 Regarding all quantities as vectors, the result is implied by the equivalence of the following relations:

(1) $|AB|^2 + |CD|^2 = |AD|^2 + |BC|^2$

(2) $|AB|^2 - |AD|^2 = |BC|^2 - |CD|^2$

(3) $(AB - AD) \cdot (AB + AD) = (BC + CD) \cdot (BC - CD)$

(4) $DB \cdot (AB + AD) = BD \cdot (BC - CD)$

(5) $DB \cdot (AB + BC + AD - CD) = 0$

(6) $2BD \cdot AC = 0$

(7) $BD \perp AC$.

2.7 Noting that

$$(g \circ f_t)\left(\frac{ax + b}{cx + d}\right) = \frac{cx + d}{(a + ct)x + b + dt},$$

we find by composing four functions of the form $g \circ f_t$ together that

$$x = \frac{1 \cdot x + 0}{0 \cdot x + 1} \rightarrow \frac{0 \cdot x + 1}{x + t_1} \rightarrow \frac{x + t_1}{t_2 x + 1 + t_1 t_2} \rightarrow$$

$$\frac{t_2 x + 1 + t_1 t_2}{(1 + t_2 t_3)x + t_1 + t_3 + t_1 t_2 t_3} \rightarrow$$

$$\frac{(1 + t_2 t_3)x + t_1 + t_3 + t_1 t_2 t_3}{(t_2 + t_4 + t_2 t_3 t_4)x + 1 + t_1 t_2 + t_1 t_4 + t_3 t_4 + t_1 t_2 t_3 t_4},$$

which will produce $2x$ provided $t_2 t_3 t_4 = -t_2 - t_4$, $t_1 t_2 t_3 = -t_1 - t_3$, and $t_2 t_3 = 1 + 2 t_3 t_4$.

This system has many solutions, one of which is $t_1 = -1$, $t_2 = \frac{2-\sqrt{2}}{2}$, $t_3 = \sqrt{2}$, $t_4 = \frac{1-\sqrt{2}}{2}$, obtained by setting $t_1 = -1$, then solving for the other values.

2.8 Note that for each A_j, if $A_j \subseteq X \subseteq S$, then X meets all of the A_i so $X \in \{A_1, \ldots, A_k\}$. Also, for $D \subseteq S$, if $D \notin \{A_1, \ldots, A_k\}$, then there exists j such that $D \cap A_j = \emptyset$, which implies $A_j \subseteq S \setminus D$ and thus $S \setminus D \in \{A_1, \ldots, A_k\}$.

Therefore, the number of subsets D of S not in $\{A_1, \ldots, A_k\}$ is k; hence $k + k = 2^n$ or $k = 2^{n-1}$.

2.9 We disprove the statement by providing a counterexample. Let the courses be designated 1, 2, 3, 4, 5, and 6. There are $\binom{6}{3} = 20$ three-element subsets of these courses. Let each of the twenty students choose a different one of these subsets. Consider two of these courses, say, without loss of generality, 1 and 2. Exactly four three-element subsets contain these courses ($\{1,2,3\}$, $\{1,2,4\}$, $\{1,2,5\}$, $\{1,2,6\}$), and exactly four three-element subsets omit the courses ($\{3,4,5\}$, $\{3,4,6\}$, $\{3,5,6\}$, $\{4,5,6\}$).

2.10 If $\sqrt{2}^{\sqrt{2}}$ is rational, then we are done (let $x = \sqrt{2}$). If $\sqrt{2}^{\sqrt{2}}$ is irrational, let $x = \sqrt{2}^{\sqrt{2}}$. Then $x^{\sqrt{2}} = 2$, and we are done.

2.11 First we note that $f'(x) = 2ax + b$ is a monotonic function which takes extreme values at $2a + b$ and $-2a + b$. Now

$$a = \frac{1}{2}\left(f(1) + f(-1) - 2f(0)\right) \quad \text{and} \quad b = \frac{1}{2}\left(f(1) - f(-1)\right),$$

so that

$$|2a + b| = \left|\frac{3}{2}f(1) + \frac{1}{2}f(-1) - 2f(0)\right| \le \frac{3}{2} + \frac{1}{2} + 2 = 4$$

and

$$|-2a + b| = \left|-\frac{1}{2}f(1) - \frac{3}{2}f(-1) + 2f(0)\right| \le \frac{1}{2} + \frac{3}{2} + 2 = 4.$$

Hence $|f'(x)| \le 4$ for all $|x| \le 1$.

Note: In general, one can show that if f is a polynomial of degree n such that $|f(x)| \le 1$ for all $|x| \le 1$, then $f'(x) \le n^2$ for all $|x| \le 1$. This result is known as Markov's theorem.

2.12 Let $w_i = v_i - v_n$, for $i = 1, \ldots, n-1$. We will show that the w_i are linearly independent and therefore $n - 1 \le d$. Note that $|w_i| = 1$ for each i, and for $i \ne j$, $|w_i - w_j| = |v_i - v_j| = 1$. Hence $w_i \cdot w_i = 1$ and for $i \ne j$, the vectors w_i, w_j, and $w_i - w_j$ form an equilateral triangle with sides of length 1, which implies that $w_i \cdot w_j = 1/2$. Suppose that

$$\alpha_1 w_1 + \cdots + \alpha_{n-1} w_{n-1} = 0.$$

Then taking the dot product of both sides of this this equation with w_1 yields the relation

$$\alpha_1 + \frac{1}{2}\alpha_2 + \cdots + \frac{1}{2}\alpha_i + \cdots + \frac{1}{2}\alpha_{n-1} = 0,$$

and for $2 \le i \le n-1$, dot producting with w_i gives

$$\frac{1}{2}\alpha_1 + \frac{1}{2}\alpha_2 + \cdots + \frac{1}{2}\alpha_{i-1} + \alpha_i + \frac{1}{2}\alpha_{i+1} + \cdots + \frac{1}{2}\alpha_{n-1} = 0.$$

Subtracting these equations implies that $\frac{1}{2}\alpha_1 - \frac{1}{2}\alpha_i = 0$, i.e., $\alpha_i = \alpha_1$. Hence all of the α_i are equal and thus all equal to 0. Therefore, the w_i are linearly independent and the result follows.

Additional Problems

2.13 Five people, Alpha, Beta, Gamma, Delta, and Epsilon, are to receive five prizes, 1, 2, 3, 4, and 5. They plan to distribute the prizes with the help of the following picture. Vertical lines are drawn from the five people's names to the five prizes, and a number of horizontal lines are placed arbitrarily between consecutive vertical lines. The only stipulation is that no two horizontal lines meet.

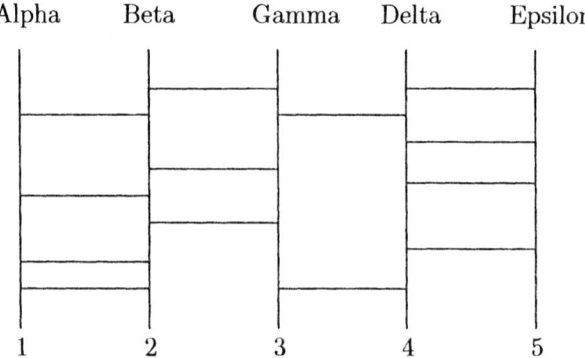

Starting with Alpha, a path is traced down the first vertical line until a horizontal line is met, the path follows the horizontal line until the next vertical line is met, and then the path moves downward. This process continues until the path ends at one of the prizes and this prize is given to Alpha. In the picture below, Alpha gets prize 2. A path like this is constructed for each of the five people. Show that no two people ever receive the same prize.

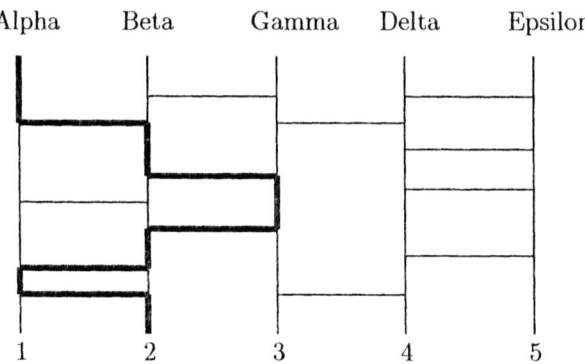

2.14 (a) You are putting together a jigsaw puzzle with 1000 pieces. Say that a "move" consists of joining any two fragments of the puzzle (a single piece qualifies as a fragment). What is the least number of moves required to complete the puzzle?

(b) A single-elimination tournament is to be held with 1000 contestants. What is the least number of matches required to complete the tournament?

2.15 I am standing somewhere on the Earth. If I go north one mile, then east one mile, and then south one mile, I end up exactly where I started. Where am I?

Hint: There is more than one solution.

2.16 Show that

$$ae + bg = cf + dh = 1 \quad \text{and} \quad af + bh = ce + dg = 0$$

if and only if

$$ae + cf = bg + dh = 1 \quad \text{and} \quad be + df = ag + ch = 0.$$

Hint: Write the given systems of equations as

$$\begin{bmatrix} a & b \\ c & d \end{bmatrix} \begin{bmatrix} e & f \\ g & h \end{bmatrix} = \begin{bmatrix} 1 & 0 \\ 0 & 1 \end{bmatrix}$$

and

$$\begin{bmatrix} e & f \\ g & h \end{bmatrix} \begin{bmatrix} a & b \\ c & d \end{bmatrix} = \begin{bmatrix} 1 & 0 \\ 0 & 1 \end{bmatrix}.$$

2.17 (Putnam Competition, 1995) Evaluate

$$\sqrt[8]{2207 - \cfrac{1}{2207 - \cfrac{1}{2207 - \ldots}}}.$$

Express your answer in the form

$$\frac{a + b\sqrt{c}}{d},$$

where a, b, c, d are integers.

Hint: Let

$$x = \sqrt[8]{2207 - \cfrac{1}{2207 - \cfrac{1}{2207 - \ldots}}}.$$

Then

$$x^8 = 2207 - \cfrac{1}{2207 - \cfrac{1}{2207 - \ldots}}$$

and therefore

$$x^8 = 2207 - \frac{1}{x^8}.$$

2.18 Alpha and Beta play a game in which they alternately write down binary digits. Thus (at the end of time) a real number in $[0, 1]$ is produced. Alpha wins if the number is transcendental and Beta wins if it is not. Alpha has a winning strategy in this game. What is it?

2.19 (Adapted from a "brain-teaser" in [4].)

Alpha thinks of an integer between 10 and 1000 (inclusive) and Beta tries to guess it.

Beta: "Is your number less than 500?"

Alpha answers the question but lies.

Beta: "Is your number a perfect square?"

Alpha answers the question but lies.

Beta: "Is your number a perfect cube?"

Alpha answers the question truthfully.

Beta: "I have narrowed your number down to two possibilities."

What is Alpha's number?

2.20 (U. S. A. Olympiad, 1981; modified) The measure of a given angle is π/n where n is a positive integer not divisible by 3. Prove that the angle can be trisected by Euclidean means (straightedge and compass).

2.21 Let
$$s(n) = 1^2 + 2^2 + \cdots + n^2.$$
Find a value of $n > 1$ for which $s(n)$ is perfect square and prove that it is the only such value.

2.22 Given five distinct points in the plane, no three collinear, show that four may be chosen which form the vertices of a convex quadrilateral.

2.23 (Putnam Competition, 1980) Let b and c be fixed real numbers and let the ten points (j, y_j), $j = 1, 2, \ldots, 10$, lie on the parabola $y = x^2 + bx + c$. For $j = 1, 2, \ldots, 9$, let I_j be the point of intersection of the tangents to the given parabola at (j, y_j) and $(j + 1, y_{j+1})$. Determine the polynomial function $y = g(x)$ of least degree whose graph passes through all nine points I_j.

2.24 Five married couples (including a host and hostess) attend a party together. At the end of the party, the host asks each of the nine others how many people he or she met for the first time that evening. He receives nine different answers. What did the hostess answer?

2.25 (*The Pentagon*, Problem 418, Fall 1988; modified) Consider the sets $\{1\}$, $\{4, 9, 16\}$, $\{25, 36, 49, 64, 81\}$, $\{100, 121, 144, 169, 196, 225, 256\}$, $\ldots$, in which each set contains two more consecutive squares than the preceding set. Find a formula for the sum of the members of the nth set.

2.26 The floor plan of rather unusual room calls for walls in the shape of a smooth simple closed curve. The room is to be illuminated by a single light placed somewhere in its interior. The ceiling and floor will not reflect light but the walls will reflect light very well. When a ray of light hits a wall it is reflected in the usual way (angle of incidence equals angle of reflection). Prove or disprove: The room can be built so that some part of it remains completely dark.

2.27 (Canadian Olympiad, 1981; modified) Let P and Q be two polynomials that satisfy the identity

$$P(Q(x)) = Q(P(x))$$

for all real numbers x. If the equation $P(x) = Q(x)$ has no real solution, show that the equation

$$P(P(x)) = Q(Q(x))$$

also has no real solution.

2.28 (Putnam Competition, 1980) Let r and s be positive integers. Derive a formula for the number of ordered quadruples (a, b, c, d) of positive integers such that

$$3^r \cdot 7^s = \text{lcm}[a, b, c] = \text{lcm}[a, b, d] = \text{lcm}[a, c, d] = \text{lcm}[b, c, d].$$

The answer should be a function of r and s.

(Note that $\text{lcm}[x, y, z]$ denotes the least common multiple of x, y, and z.)

2.29 (T. Hamlin) A bin contains n balls, numbered $1, \ldots, n$. Alpha selects n_1 balls at random from the bin, replaces the balls, and then Beta selects n_2 balls at random. Show that the expected number of matches between the two sets chosen is

$$\frac{n_1 n_2}{n}.$$

(A match means that the same ball is chosen by both Alpha and Beta.)

2.30 Three people, Alpha, Beta, and Gamma, have nonnegative integers on their foreheads, so that they can see each other's numbers but not their own. Three numbers are written on the blackboard, and the people are told that one of the numbers on the board is the sum of the numbers on their foreheads. The three people are asked, in cyclical order, the question, "Do you know your number?" Show that eventually someone says, "Yes."

Chapter 3

Contradiction

Test an absurdity and you may stumble on a truth.

ROY C. VICKERS
Detective-Inspector Rason,
"The Case of the Social Climber,"
The Department of Dead Ends, 1978

If you suspect that a proposition is true, assume that it is false and see if a contradiction results. If so, then you have proved that the proposition is indeed true.

Let us recall the most famous proof by contradiction of all time:

Example 3.1. Show that $\sqrt{2}$ is an irrational number.

Solution: Suppose that $\sqrt{2}$ is a rational number. Then

$$\sqrt{2} = \frac{m}{n},$$

where m and n are relatively prime positive integers. It follows that

$$2 = \frac{m^2}{n^2},$$

and $2n^2 = m^2$. Hence m^2 is an even number and therefore m itself is even. Let $m = 2k$. Then $2n^2 = 4k^2$ and therefore $n^2 = 2k^2$. Hence n is an even integer. However, we assumed that m and n are relatively prime and therefore *not* both even. This contradiction shows that the assumption that $\sqrt{2}$ is a rational number is false. Therefore, $\sqrt{2}$ is an irrational number. ■

It is not much more difficult to show that e is irrational.

Example 3.2. Prove that e is an irrational number.

Solution: Recall that

$$e = \frac{1}{0!} + \frac{1}{1!} + \frac{1}{2!} + \frac{1}{3!} + \cdots.$$

Suppose that e is a rational number, say, $e = p/q$. We note that $q \neq 1$, for

$$2 < e < 2 + \frac{1}{2!}\left(1 + \frac{1}{3} + \frac{1}{3^2} + \cdots\right) = 2 + \frac{1}{2}\left(\frac{3}{2}\right) = 2 + \frac{3}{4} < 3.$$

Now

$$\frac{p}{q} = \frac{1}{0!} + \frac{1}{1!} + \frac{1}{2!} + \cdots$$

$$= \left(\frac{1}{0!} + \cdots + \frac{1}{q!}\right) + \left(\frac{1}{(q+1)!} + \cdots\right),$$

and hence

$$q!\frac{p}{q} = q!\left(\frac{1}{0!} + \cdots + \frac{1}{q!}\right) + q!\left(\frac{1}{(q+1)!} + \cdots\right).$$

The expression on the left side is an integer and the first term on the right side is an integer. But the second term on the right side is a number between 0 and 1:

$$0 < q!\left(\frac{1}{(q+1)!} + \cdots\right) < \frac{1}{q} + \frac{1}{q^2} + \frac{1}{q^3} + \cdots = \frac{1}{q-1} \leq 1.$$

This is a contradiction. Hence e is an irrational number.

Note: Similarly, one can show that

$$x = \frac{1}{1!} + \frac{1}{1!\,2!} + \frac{1}{1!\,2!\,3!} + \frac{1}{1!\,2!\,3!\,4!} + \cdots$$

is an irrational number. See Additional Problem 3.11.   ∎

Here is a Putnam problem easily solved by proof by contradiction:

Example 3.3. (Putnam Competition, 1991) Let A and B be different $n \times n$ matrices with real entries. If $A^3 = B^3$ and $A^2B = B^2A$, can $A^2 + B^2$ be invertible?

Solution: No. Suppose that $A^2 + B^2$ is invertible. Then $C(A^2 + B^2) = I$ for some matrix C. It follows that $C(A^3 + B^2 A) = A$ and $C(A^2 B + B^3) = B$. But then $A = C(A^3 + B^2 A) = C(B^3 + A^2 B) = B$, contradicting the assumption that A and B are different matrices. Therefore $A^2 + B^2$ is not invertible. ∎

The next problem provides a relatively bizarre example of the "reductio ad absurdum" technique.

Example 3.4. Suppose that $v_1, \ldots, v_n$ are points in $\mathbf{R}^d$ with the property that $|v_i - v_j| = 1$ for $i \neq j$. Show that $n \leq d + 1$.

Solution: Note that this is Problem 2.12 which was solved by defining $w_i = v_i - v_n$ for $i = 1, \ldots, n - 1$ and then showing (directly) that the w_i are linearly independent. Here we will show this by contradiction.

So suppose that the w_i are linearly dependent. Then some one of them can be expressed as a linear combination of the others. Without loss of generality, assume that $w_1 = \sum_{i=2}^{n-1} t_i w_i$. Taking the dot product of this equation with w_1, and recalling that $w_i \cdot w_i = 1$ and $w_i \cdot w_j = 1/2$ for $i \neq j$, we find that $\sum_{i=2}^{n-1} t_i = 2$. But if we dot product with w_j for $j \geq 2$, we obtain

$$\frac{1}{2} = t_j + \frac{1}{2} \sum_{i \neq j} t_i,$$

which implies each $t_j = -1$! ∎

We conclude with one more problem about irrational numbers.

Example 3.5. Without using countability arguments, prove that there exists x irrational such that x^x is irrational.

Solution: Define x by the equation $x^x = \pi$. (Such an x exists by the intermediate value theorem, since $f(x) = x^x$ is a continuous function, $f(1) = 1$, and $f(2) = 4$.) Suppose that x is rational, say, $x = m/n$. Then

$$\left(\frac{m}{n}\right)^{m/n} = \pi,$$

so that

$$n^m \pi^n = m^m.$$

But this relation contradicts the fact that π is a transcendental number (it does not satisfy any polynomial equation with integer coefficients). Hence both x and x^x are irrational. ∎

Problems

3.1 (a) Prove that there are no integers $x > 1$, $y > 1$, $z > 1$ with

$$x! + y! = z!.$$

(b) Find infinitely many integers $x > 1$, $y > 1$, $z > 1$ with

$$x!y! = z!.$$

3.2 Prove that there do not exist positive integers x, y, z such that

$$x^n + y^n = z^n$$

for all integers $n \geq 3$.

Note: This result is much weaker than Fermat's (last) theorem, which asserts that the above equation has a solution in integers x, y z for *no $n \geq 3$*.

3.3 Prove that it is impossible for the decimal expansions of e and π to contain each other.

3.4 Prove that there is at least one nonzero digit between the millionth and the three-millionth decimal digits of $\sqrt{2}$.

3.5 Define an equivalence relation on the set of real numbers by saying that two real numbers x and y are equivalent if and only if $x - y$ is a rational number. Show that there are uncountably many equivalence classes.

3.6 Let A be a finite set of integers. Let $r(n)$ denote the number of ways of writing n in the form $a - a'$, where a, $a' \in A$. Show that $\max r(n) = r(0)$.

3.7 Let α and β be positive irrational numbers with $1/\alpha + 1/\beta = 1$. Prove that the integers $[\alpha]$, $[\beta]$, $[2\alpha]$, $[2\beta]$, $[3\alpha]$, $[3\beta]$, ... are a permutation of 1, 2, 3, ...

3.8 Show that there is no nonconstant polynomial f with integral coefficients for which $f(k)$ is a prime number for every positive integer k.

3.9 If n^n has 10^{100} decimal digits, how many has n?

3.10 Prove that $\tan(\pi/3n)$ is irrational for all positive integers n.

Solutions

3.1 (a) Suppose that $x! + y! = z!$. Clearly $z > x$ and $z > y$. Since $x! | x!$ and $x! | z!$, it follows that $x! | y!$. Similarly, $y! | x!$. Thus $x! = y!$ and $x = y$. But this implies that $2x! = z!$, which is impossible, since

$$2x! < (x + 1)x! = (x + 1)! \leq z!.$$

(b) For $y \geq 3$, let $x = y! - 1$ and $z = y!$. Then

$$x! y! = (y! - 1)! y! = (y!)! = z!.$$

Hence the equation has infinitely many solutions.

3.2 Suppose that there exist positive integers x, y, z such that $x^n + y^n = z^n$ for all integers $n \geq 3$. Since x and y are positive, it follows that z is greater than both x and y. Without loss of generality, suppose that $x \geq y$. Then

$$z^n = x^n + y^n \leq 2x^n \leq 2(z - 1)^n.$$

However, the inequality $z^n \leq 2(z - 1)^n$ fails when n is sufficiently large. Therefore our original assumption is false, and there are no integers x, y, z such that $x^n + y^n = z^n$ for all integers $n \geq 3$.

Note: The equations $x^n + y^n = z^n$ and $x^{n+1} + y^{n+1} = z^{n+1}$ never both hold for the same values of x, y, z. For, multiplying the first equation by z, we obtain

$$z \cdot x^n + z \cdot y^n = z^{n+1} = x^{n+1} + y^{n+1} = x \cdot x^n + y \cdot y^n,$$

and hence $x^n(z - x) = y^n(y - z)$. This equation does not hold since $z - x$ is positive and $y - z$ is negative.

3.3 Solution (1): Suppose that the decimal expansions of e and π contain each other. Suppose that the digits of π occur within the expansion of e beginning at the ath digit and the digits of e occur within the expansion of π beginning at the bth digit. Then both e and π have periodic decimal expansions, with period $a + b - 2$, contradicting the fact that they are irrational numbers.

Solution (2): Again, suppose that the decimal expansions of e and π contain each other. Then $10^m \pi - e = q_1$ and $10^n e - \pi = q_2$, for integers m, n, q_1, and q_2. Multiplying the first equation by 10^n and adding the result to the second equation, we find that $(10^{m+n} - 1)\pi = 10^n q_1 + q_2$, contradicting the fact that π is an irrational number.

3.4 Suppose that $\sqrt{2}$ contains no nonzero digit between the millionth and three-millionth digits of its decimal expansion. Then

$$10^{10^6}\sqrt{2} = I + r,$$

where I is a positive integer, $I < \sqrt{2} \cdot 10^{10^6}$, and $0 < r < 10^{-2 \cdot 10^6}$. It follows that

$$2 \cdot 10^{2 \cdot 10^6} = I^2 + r^2 + 2Ir,$$

and $r^2 + 2Ir$ is required to be an integer. But

$$0 < r^2 + 2Ir < 10^{-4 \cdot 10^6} + 2\sqrt{2} \cdot 10^{10^6} 10^{-2 \cdot 10^6} < 1,$$

a contradiction.

3.5 First we show that each equivalence class has a countable number of elements. Suppose that C is an equivalence class and $x \in C$. Since $y \in C$ if and only if $x - y$ is rational, there are countably many $y \in C$. In other words, C has countably many elements. (Note that $x - y_1 = x - y_2$ only if $y_1 = y_2$.) Now, if there were only countably many equivalence classes, then $\mathbf{R}$ would be the union of countably many countable sets, which is false. Therefore there are uncountably many equivalence classes.

3.6 Clearly, $\max r(n) \geq r(0)$. We just need to show that $\max r(n) \leq r(0)$. Evidently, $r(0) = |A|$. Suppose that $\max r(n) > |A|$. Then

$$n = a_1 - a_1' = a_2 - a_2' = \cdots = a_{|A|+k} - a_{|A|+k}',$$

and some $a_i = a_j$, which implies that $a_i' = a_j'$, and the two representations $a_i - a_i'$ and $a_j - a_j'$ are not distinct.

3.7 Since α and β are irrational, $m\alpha$ and $n\beta$ are never integers. Furthermore, since $\alpha > 1$, $[m\alpha] \neq [(m+1)\alpha]$ for any m; the same is true for β. Now we want to show two things:

(1) For no m, n do we have $[m\alpha] = [n\beta]$.

(2) For no N do we have $[m\alpha] < N$, $[\alpha(m+1)] > N$, $[n\beta] < N$, and $[(n+1)\beta] > N$.

Proof of (1): Suppose that $[m\alpha] = [n\beta]$. Then $m\alpha - 1 < N < m\alpha$ and $n\beta - 1 < N < n\beta$ for some integer N. It follows that $m/N > 1/\alpha$ and $n/N > 1/\beta$. Adding these two inequalities yields $m + n > N$. However, we also obtain the inequalities $1/\alpha > m/(N+1)$ and $1/\beta > n/(N+1)$ and adding these two inequalities yields $N + 1 > m + n$. Hence $N < m + n < N + 1$, a contradiction since $m + n$ is an integer.

Proof of (2): Suppose that $[m\alpha] < N$, $[(m+1)\alpha] > N$, $[n\beta] < N$, and $[(n+1)\beta] > N$. Then $m\alpha < N$, $n\beta < N$, $N+1 < (m+1)\alpha$, and $N+1 < (n+1)\beta$. It follows that $m/N < 1/\alpha$ and $n/N < 1/\beta$. Adding these two inequalities yields $m+n < N$. It also follows that $1/\alpha < (m+1)/(N+1)$ and $1/\beta < 1/(n+1)/(N+1)$. Adding these two inequalities yields the inequality $N+1 < m+n+2$. From these inequalities it follows that $m+n < N < m+n+1$, contradicting the fact that N is an integer.

3.8 Suppose that there is such a polynomial

$$f(x) = a_n x^n + \cdots + a_0,$$

with integral coefficients. Suppose that $f(1) = p$. Then, for all m,

$$
\begin{aligned}
f(1+mp) - f(1) &= a_n(1+mp)^n + \cdots + a_0 - a_n - \cdots - a_0 \\
&= a_n mp + \cdots + a_n m^n p^n + \cdots + mpa_1,
\end{aligned}
$$

which is divisible by p. Hence $f(1+mp)$ is divisible by p for all m. In order that $f(k)$ be always prime it is necessary that $f(1+mp) = p$ for every m. But no nonconstant polynomial takes on the same value infinitely often; we have a contradiction to the hypothesis that $f(k)$ is always prime.

3.9 We have $10^{10^{100}-1} \le n^n < 10^{10^{100}}$. Let k be the number of decimal digits of n. Then $10^{k-1} \le n < 10^k$. Suppose that $k > 99$. Then $k \ge 100$ which implies $n \ge 10^{99}$ and hence

$$n^n \ge (10^{99})^{10^{99}} = 10^{99\cdot 10^{99}} > 10^{10^{100}},$$

a contradiction. Therefore, $k \le 99$.

Now suppose that $k < 99$. Then $k \le 98$ which implies $n < 10^k \le 10^{98}$ and hence

$$n^n < (10^{98})^{10^{98}} = 10^{(100-2)\cdot 10^{98}} = 10^{10^{100}-2\cdot 10^{98}} < 10^{10^{100}-1},$$

another contradiction. Therefore $k = 99$.

3.10 Assume that $\tan(\pi/3n)$ is rational for some $n \ge 1$. We will show by induction that $\tan(\pi m/3n)$ is rational for all $m \ge 1$. The base case is our assumption that $\tan(\pi/3n)$ is rational. Assume that $\tan(\pi k/3n)$ is rational. Then

$$\tan\frac{\pi(k+1)}{3n} = \frac{\tan\frac{\pi k}{3n} + \tan\frac{\pi}{3n}}{1 - \tan\frac{\pi k}{3n}\tan\frac{\pi}{3n}}$$

is rational. Thus $\tan(\pi m/3n)$ is rational for all $m \ge 1$. But letting $m = n$ we conclude that $\tan \pi/3$ is rational, contradicting the fact that $\tan \pi/3 = \sqrt{3}$. Hence $\tan(\pi/3n)$ is irrational for all n.

Additional Problems

3.11 Show that

$$x = \frac{1}{1!} + \frac{1}{1!\,2!} + \frac{1}{1!\,2!\,3!} + \frac{1}{1!\,2!\,3!\,4!} + \cdots$$

is an irrational number.

3.12 (Putnam Competition, 1955; modified) Prove that no integers m, n, and p, not all 0, satisfy $m + n\sqrt{2} + p\sqrt{3} = 0$.

3.13 (Putnam Competition, 1975) Does there exist a subset B of the unit circle $x^2 + y^2 = 1$ such that (i) B is topologically closed, and (ii) B contains exactly one point from each pair of diametrically opposite points on the circle? [A set B is *topologically closed* if it contains the limit point of every convergent sequence of points in B.]

3.14 Prove that the logarithm function, $y = \log x$, is not algebraic. That is, y satisfies no equation of the form

$$p_n(x)y^n + p_{n-1}(x)y^{n-1} + \cdots + p_1(x)y + p_0(x) = 0,$$

where the p_i are polynomials in x with $p_n \neq 0$.

3.15 Let $P(x)$ be a polynomial in x with real coefficients of degree $n \geq 2$ whose first three terms are $x^n + 2x^{n-1} + 2x^{n-2}$. Show that $P(x)$ cannot have n distinct real roots.

3.16 (Canadian Olympiad, 1981; modified) For any real number t denote by $[t]$ the greatest integer which is less than or equal to t. Show that the equation,

$$[x] + [2x] + [4x] + [8x] + [16x] + [32x] = 12345$$

has no real solution.

3.17 Show that for each positive integer n, there exists a circle containing exactly n lattice points in its interior.

3.18 (Putnam Competition, 1981) Two distinct squares of the 8 by 8 chessboard C are said to be adjacent if they have a vertex or side in common. Also, g is called a C-gap if for every numbering of the squares with all the integers 1, 2, ..., 64 there exist two adjacent squares whose numbers differ by at least g. Determine the largest C-gap g.

3.19 Prove that there are no integers m, n, k, all greater than 1, such that $n! = m^k$.

3.20 Show that the sum of an uncountable number of positive quantities is never a finite number.

3.21 Is it possible to place uncountably many nonoverlapping (pairwise disjoint) copies of the letter A in the plane? (Assume that the A is made of line segments with no width.) The A's may be translated, rotated, shrunk, or expanded. Of which capital letters, A, B, ..., Z, is it possible to place uncountably many copies in the plane without overlapping?

3.22 Let

$$X = \{(x, y, z) \in \mathbf{R}^3 : 0 \le x, \, y \le 1, \, z = 0\}$$
$$Y = \{(x, y, z) \in \mathbf{R}^3 : x = 0, \, y = 0, \, 0 \le z \le 1\}$$
$$Z = \{(x, y, z) \in \mathbf{R}^3 : x = 1/2, \, y = 1/2, \, 0 \le z \le 1\}.$$

Is it possible to place uncountably many nonoverlapping copies of the set $X \cup Y$ in $\mathbf{R}^3$? Is the answer the same for the set $X \cup Z$? As in the previous problem, the sets may be translated, rotated, shrunk, or expanded.

3.23 (a) Let S be a set of countably many points in the plane. Show that there exist uncountably many nonoverlapping translates of S.

(b) Let S be a union of countably many lines in $\mathbf{R}^3$. Show that there exist uncountably many nonoverlapping translates of S.

Chapter 4

Induction

"It's been my experience that secrets are hard to keep. If fathers
know and sons know, pretty soon other people know."

TONY HILLERMAN
Lt. Joe Leaphorn, *Listening Woman*, 1978.

A linear order relation for a set A is called a "well-order" relation for A
provided every nonempty subset of A contains a smallest element.

The usual $\leq$ ordering for the set $\mathbf{N}$ of all positive integers is an example
of a well-order relation. A consequence of this fact is the following principle
of mathematical induction:

Theorem 4.1 (Principle of mathematical induction). If a subset M
of $\mathbf{N}$ satisfies the two conditions, (i) $1 \in M$ and (ii) $n \in M$ implies that
$n + 1 \in M$, then necessarily $M = \mathbf{N}$.

Proof. The proof of this theorem is by contradiction. If there exists any
positive integer not belonging to M, then (since $\mathbf{N}$ is well-ordered) there
must be a smallest such integer; call it m. Now (i) implies that $m \neq 1$;
hence $m > 1$ and so $m - 1$ is a positive integer. Because of (ii), $m - 1$
doesn't belong to M. But this contradicts m being the smallest integer
with this property. ∎

Induction can frequently be used to solve problems if the problem solution involves establishing that a certain result is true for all positive integers.
First show that the result holds for the positive integer 1. Then show that
the result holds for $n + 1$ under the assumption that it holds for n. The
principle of mathematical induction then guarantees the result is valid for
all positive integers.

Slight variations of the induction principle frequently occur. For example, if a result is perhaps false for $n < 4$ but is true for $n = 4$ and true for $n + 2$ whenever it is true for n, the conclusion is that the result holds for all even integers greater than or equal to 4.

Sometimes the conclusion that a result holds for $n + 1$ does not follow from just the assumption that it holds for n, but does follow from the assumption that it holds for all positive integers less than $n + 1$. For problems of this type, a modified induction principle is valid: If $M \subseteq \mathbf{N}$ satisfies (i) $1 \in M$ and (ii) $\{1, 2, \ldots, n\} \subseteq M$ implies $n + 1 \in M$, then $M = \mathbf{N}$.

Condition (i) in the statement of the theorem is often called the *basis* of the induction, and the premise of condition (ii) is usually called the *induction hypothesis* or the *induction assumption*.

Here are some examples illustrating the use of mathematical induction in problem solving.

Example 4.1. Prove that $\displaystyle\sum_{k=1}^{n} \frac{1}{\sqrt{k}} < 2\sqrt{n}$.

Solution: We Prove the result by induction on n. For $n = 1$, the inequality asserts that $1 < 2$, which is true.

If $\displaystyle\sum_{k=1}^{n} \frac{1}{\sqrt{k}} < 2\sqrt{n}$ for some $n \geq 1$, then

$$\sum_{k=1}^{n+1} \frac{1}{\sqrt{k}} = \sum_{k=1}^{n} \frac{1}{\sqrt{k}} + \frac{1}{\sqrt{n+1}} < 2\sqrt{n} + \frac{1}{\sqrt{n+1}} = \frac{2\sqrt{n^2+n}+1}{\sqrt{n+1}}$$

$$< \frac{2\sqrt{n^2+n+\frac{1}{4}}+1}{\sqrt{n+1}} = \frac{2(n+\frac{1}{2})+1}{\sqrt{n+1}} = \frac{2(n+1)}{\sqrt{n+1}} = 2\sqrt{n+1},$$

and hence the inequality holds for all positive integers n.   ∎

Example 4.2. Show that $5^n + 6 \cdot 7^n + 1$ is divisible by 8 for all nonnegative integers n.

Solution: The result is clear if $n = 0$. Assume that $5^n + 6 \cdot 7^n + 1$ is divisible by 8, say, $5^n + 6 \cdot 7^n + 1 = 8k$ where k is an integer. Then

$$5^{n+1} + 6 \cdot 7^{n+1} + 1 = 5 \cdot 5^n + 42 \cdot 7^n + 1$$

$$= 5(5^n + 6 \cdot 7^n + 1) + 12 \cdot 7^n - 4$$

$$= 40k + 4(3 \cdot 7^n - 1),$$

and since $3 \cdot 7^n - 1$ is an even integer, it follows that $5^{n+1} + 6 \cdot 7^{n+1} + 1$ is also divisible by 8. ∎

Example 4.3. Use induction to establish Abel's formula:

$$\sum_{k=1}^{n} a_k b_k = A_n b_n - \sum_{k=1}^{n-1} A_k (b_{k+1} - b_k)$$

for $n \geq 2$, where $A_k = \sum_{i=1}^{k} a_i$.

Solution: For $n = 2$, the right side is

$$A_2 b_2 - A_1(b_2 - b_1) = (a_1 + a_2)b_2 - a_1(b_2 - b_1) = a_1 b_1 + a_2 b_2,$$

so the formula holds for $n = 2$.

Assuming that the formula is correct for n, then

$$A_{n+1}b_{n+1} - \sum_{k=1}^{n} A_k(b_{k+1} - b_k) = (A_n + a_{n+1})b_{n+1}$$

$$-\sum_{k=1}^{n-1} A_k(b_{k+1} - b_k) - A_n(b_{n+1} - b_n)$$

$$= A_n b_n - \sum_{k=1}^{n-1} A_k(b_{k+1} - b_k) + a_{n+1}b_{n+1}$$

$$= \sum_{k=1}^{n} a_k b_k + a_{n+1}b_{n+1}$$

$$= \sum_{k=1}^{n+1} a_k b_k.$$

∎

Example 4.4. A subset A of a real vector space V is called a *convex set* provided that $tx + (1-t)y \in A$ whenever x and $y \in A$ and $0 < t < 1$. If A is convex, show that $\sum_{j=1}^{n} t_j x_j \in A$ whenever each $x_j \in A$ and each $t_j > 0$ with $\sum_{j=1}^{n} t_j = 1$.

Solution: The result is trivially true if $n = 1$, and is the definition of convexity if $n = 2$. Assume that the result is true for some $n \geq 2$ and suppose that $x_1, x_2, \ldots, x_n, x_{n+1} \in A$ and each $t_j > 0$ with $\displaystyle\sum_{j=1}^{n+1} t_j = 1$. Then

$$\sum_{j=1}^{n+1} t_j x_j = \sum_{j=1}^{n} t_j x_j + t_{n+1} x_{n+1} = (1 - t_{n+1}) \sum_{j=1}^{n} \frac{t_j}{1 - t_{n+1}} x_j + t_{n+1} x_{n+1}.$$

Now each $t_j/(1 - t_{n+1}) > 0$ and

$$\sum_{j=1}^{n} \frac{t_j}{1 - t_{n+1}} = \frac{1}{1 - t_{n+1}} \sum_{j=1}^{n} t_j = \frac{1}{1 - t_{n+1}}(1 - t_{n+1}) = 1,$$

so by the induction assumption, $x \equiv \displaystyle\sum_{j=1}^{n} \frac{t_j}{1 - t_{n+1}} x_j \in A$ and hence

$$\sum_{j=1}^{n+1} t_j x_j = (1 - t_{n+1})x + t_{n+1} x_{n+1} \in A \text{ since } A \text{ is convex.} \qquad \blacksquare$$

Example 4.5. Suppose that $a_0 = 2$, $a_1 = 7$, and

$$a_k = a_{\lfloor k/3 \rfloor} + a_{\lfloor (k+1)/3 \rfloor} + a_{\lfloor (k+2)/3 \rfloor} - 4$$

for $k \geq 2$, (where $\lfloor x \rfloor$ denotes the greatest integer $\leq x$). Show that

$$a_k = 5k + 2$$

for all nonnegative integers k.

Solution: The result is clear if $k = 0$ or 1. Assume that the result is true for all nonnegative integers smaller than k.

Case 1. If $k \equiv 0 \bmod 3$, say $k = 3j$, then $5k + 2 = 15j + 2$ and also (using the recurrence formula and applying the induction assumption to j), $a_k = 3a_j - 4 = 3(5j + 2) - 4 = 15j + 2$.

Case 2. If $k = 3j + 1$, then $a_k = 2a_j + a_{j+1} - 4 = 2(5j + 2) + 5(j + 1) + 2 - 4 = 15j + 7 = 5k + 2$.

Case 3. If $k = 3j + 2$, then $a_k = a_j + 2a_{j+1} - 4 = 5j + 2 + 2[5(j + 1) + 2] - 4 = 15j + 12 = 5k + 2$. $\quad\blacksquare$

Problems

4.1 For all x in the interval $0 \leq x \leq \pi$, prove that $|\sin nx| \leq n \sin x$, where n is a nonnegative integer.

4.2 Let f_n denote the nth term in the Fibonacci sequence: $f_0 = 0$, $f_1 = 1$, and $f_{n+2} = f_{n+1} + f_n$ for $n \geq 0$. Prove that $f_{n+1}^2 + f_n^2 = f_{2n+1}$.

4.3 Let $0 < a_1 < a_2 < \cdots < a_n$, and let $e_i = \pm 1$. Prove that $\sum_{i=1}^{n} e_i a_i$ assumes at least $\binom{n+1}{2}$ distinct values as the e_i range over the 2^n possible combinations of signs.

4.4 Discover and prove a formula for

$$\prod_{i=1}^{n} \left\{ 1 - \frac{4}{(2i-1)^2} \right\}.$$

Note: This exercise also appeared as Problem 1.3.

4.5 Let $a_1, \ldots, a_n$ and $b_1, \ldots, b_n$ be positive real numbers. For each i, let m_i be the smaller of the two quantities a_i and b_i and let M_i be the larger of the two. Prove that

$$m_1 m_2 \ldots m_n + M_1 M_2 \ldots M_n \geq a_1 a_2 \ldots a_n + b_1 b_2 \ldots b_n.$$

4.6 Evaluate

$$\prod_{n=0}^{\infty} \left(1 + 2^{-2^n} \right).$$

4.7 Let $x^{(0)} = 1$, and for n a positive integer, let

$$x^{(n)} = x(x-1)\ldots(x-n+1).$$

Prove that

$$(x+y)^{(n)} = \sum_{k=0}^{n} \binom{n}{k} x^{(k)} y^{(n-k)}$$

for all $n \geq 0$.

4.8 Prove that

$$\sum_{k=1}^{n} \frac{1}{k} > \ln n$$

for $n \geq 1$.

4.9 Prove that

$$\sum_{k=1}^{n} \frac{1}{k} < \ln(2n+1)$$

for $n \geq 1$.

4.10 For the Fibonacci numbers f_n, show that

$$\sum_{i=1}^{n} f_i^3 = \frac{1}{10} f_{3n+2} + \frac{3}{5}(-1)^{n-1} f_{n-1} + \frac{1}{2}.$$

4.11 For the Fibonacci numbers f_n, prove the following statements.

(a) For $n \geq 0$, $m \geq 1$,

$$f_{n+m} = f_m f_{n+1} + f_{m-1} f_n.$$

(b) If $a|b$, then $f_a|f_b$.

(c) For $a, b \geq 1$,

$$f_{\gcd(a,b)} = \gcd(f_a, f_b).$$

4.12 If a is a real constant and if D denotes differentiation with respect to x, show that

$$(D-a)^n (x^j e^{ax}) = 0$$

for all integers n and j with $n \geq 1$ and $0 \leq j \leq n-1$.

4.13 For real numbers $r_1, r_2, \ldots, r_n$ establish the Vandermonde determinant formula $V_n = \displaystyle\prod_{1 \leq i < j \leq n} (r_j - r_i)$, where

$$V_n = \begin{vmatrix} 1 & 1 & \cdots & 1 \\ r_1 & r_2 & \cdots & r_n \\ r_1^2 & r_2^2 & \cdots & r_n^2 \\ \vdots & \vdots & \vdots & \vdots \\ r_1^{n-1} & r_2^{n-1} & \cdots & r_n^{n-1} \end{vmatrix}.$$

4.14 Prove that every positive rational number is representable as the sum of a finite number of distinct terms from the sequence $1, 1/2, 1/3, 1/4, \ldots$.

Note: Numbers of the form $1/q$ are called *Egyptian fractions* because the ancient Egyptians used them to represent all fractions (except $2/3$, for which they had a special hieroglyph).

Solutions

4.1 If $n = 0$, the result is clear since $0 \leq 0$, and the $n = 1$ case is also clear since $|\sin x| = \sin x$ for $0 \leq x \leq \pi$.

If $|\sin kx| \leq k \sin x$, for some $k \geq 1$, then

$$
\begin{aligned}
|\sin(k+1)x| &= |\sin(kx + x)| \\
&= |\sin kx \cos x + \cos kx \sin x| \\
&\leq |\sin kx| \cdot |\cos x| + |\cos kx| \cdot |\sin x| \\
&\leq |\sin kx| + \sin x \\
&\leq k \sin x + \sin x \\
&= (k+1) \sin x.
\end{aligned}
$$

4.2 For $n = 0$, $f_1^2 + f_0^2 = 1 + 0 = 1 = f_1$, and for $n = 1$, $f_2^2 + f_1^2 = 1 + 1 = 2 = f_3$.

If $f_{j+1}^2 + f_j^2 = f_{2j+1}$ for all j, $0 \leq j \leq n$ (for some $n \geq 1$), then

$$
\begin{aligned}
f_{2n+3} &= f_{2n+2} + f_{2n+1} \\
&= f_{2n+1} + f_{2n} + f_{2n+1} \\
&= 2f_{2n+1} + f_{2n+1} - f_{2n-1} \\
&= 3f_{2n+1} - f_{2n-1} \\
&= 3(f_{n+1}^2 + f_n^2) - (f_n^2 + f_{n-1}^2) \\
&= 3f_{n+1}^2 + 2f_n^2 - f_{n-1}^2 \\
&= f_{n+1}^2 + 2f_{n+1}^2 + 2f_n^2 - f_{n-1}^2 \\
&= f_{n+1}^2 + (f_{n+1} + f_n)^2 + f_{n+1}^2 + f_n^2 - 2f_{n+1}f_n - f_{n-1}^2 \\
&= f_{n+1}^2 + f_{n+2}^2 + (f_{n+1} - f_n)^2 - f_{n-1}^2 \\
&= f_{n+1}^2 + f_{n+2}^2 + f_{n-1}^2 - f_{n-1}^2 \\
&= f_{n+1}^2 + f_{n+2}^2.
\end{aligned}
$$

4.3 The result is clear if $n = 1$. Assume that the result is correct for some n and suppose that $0 < a_1 < a_2 < \cdots < a_n < a_{n+1}$. By the induction assumption, $\sum_{i=1}^{n} e_i a_i + a_{n+1}$ assumes at least $\binom{n+1}{2}$ distinct values, the smallest of which is $-\sum_{i=1}^{n} a_i + a_{n+1}$. Noting the $n+1$ additional values,

$$-\sum_{i=1}^{n+1} a_i < a_1 - \sum_{i=2}^{n+1} a_i < -a_1 + a_2 - \sum_{i=3}^{n+1} a_i < \cdots < -\sum_{i=1}^{n-1} a_i + a_n - a_{n+1},$$

all of which are smaller than $-\sum_{i=1}^{n} a_i + a_{n+1}$, it follows that $\sum_{i=1}^{n+1} e_i a_i$ assumes at least $\binom{n+1}{2} + (n+1) = \binom{n+2}{2}$ distinct values.

4.4 Let x_n be the quantity in question. Computing the first few values, $x_1 = -3/1$, $x_2 = -5/3$, $x_3 = -7/5$, suggests the formula $x_n = -(2n+1)/(2n-1)$. Assuming this to be correct for x_n, then

$$
\begin{aligned}
x_{n+1} &= \frac{-(2n+1)}{2n-1} \cdot \left[1 - \frac{4}{(2n+1)^2} \right] \\
&= \frac{-(2n+1)(2n+3)(2n-1)}{(2n-1)(2n+1)^2} \\
&= \frac{-(2n+3)}{2n+1}.
\end{aligned}
$$

4.5 For $n = 1$, we have $m_1 + M_1 = a_1 + b_1$. Assume that the result holds for n and suppose that we have $n+1$ such number pairs.

Case 1. If $a_{n+1} \le b_{n+1}$, then

$$
\begin{aligned}
m_1 \ldots m_n m_{n+1} &+ M_1 \ldots M_n M_{n+1} \\
&= m_1 \ldots m_n a_{n+1} + M_1 \ldots M_n b_{n+1} \\
&= (m_1 \ldots m_n + M_1 \ldots M_n) a_{n+1} + M_1 \ldots M_n (b_{n+1} - a_{n+1}) \\
&\ge (a_1 \ldots a_n + b_1 \ldots b_n) a_{n+1} + b_1 \ldots b_n (b_{n+1} - a_{n+1}) \\
&= a_1 \ldots a_n a_{n+1} + b_1 \ldots b_n b_{n+1}.
\end{aligned}
$$

Case 2. If $a_{n+1} > b_{n+1}$, then

$$
\begin{aligned}
m_1 \ldots m_n m_{n+1} &+ M_1 \ldots M_n M_{n+1} \\
&= m_1 \ldots m_n b_{n+1} + M_1 \ldots M_n a_{n+1} \\
&= (m_1 \ldots m_n + M_1 \ldots M_n) b_{n+1} + M_1 \ldots M_n (a_{n+1} - b_{n+1}) \\
&\ge (a_1 \ldots a_n + b_1 \ldots b_n) b_{n+1} + a_1 \ldots a_n (a_{n+1} - b_{n+1}) \\
&= a_1 \ldots a_n a_{n+1} + b_1 \ldots b_n b_{n+1}.
\end{aligned}
$$

4.6 Solution (1): Let $x_n = \prod_{j=0}^{n} (1 + 2^{-2^j})$. Then $x_0 = 3/2 = (2^2 - 1)/2^1$,

$x_1 = 15/8 = (2^4 - 1)/2^3$, $x_2 = 255/128 = (2^8 - 1)/2^7$, which suggests
the formula $x_n = \left(2^{2^{n+1}} - 1\right) / \left(2^{2^{n+1}-1}\right)$.

Assuming that this is true for some n, we have

$$x_{n+1} = \frac{2^{2^{n+1}} - 1}{2^{2^{n+1}-1}} \left(1 + 2^{-2^{n+1}}\right)$$

$$= \frac{\left(2^{2^{n+1}} - 1\right)\left(2^{2^{n+1}} + 1\right)}{2^{2^{n+1}-1} \cdot 2^{2^{n+1}}}$$

$$= \frac{2^{2^{n+2}} - 1}{2^{2 \cdot 2^{n+1}-1}}$$

$$= \frac{2^{2^{n+2}} - 1}{2^{2^{n+2}-1}}.$$

So by induction,

$$x_n = \frac{2^{2^{n+2}} - 1}{2^{2^{n+2}-1}} = 2 - \frac{1}{2^{2^{n+2}-1}},$$

which approaches 2 as n approaches infinity. Hence the value of the
infinite product is 2.

Solution (2): We have

$$\prod_{n=0}^{\infty} \left(1 + 2^{-2^n}\right) = \lim_{k \to \infty} \prod_{n=0}^{k-1} \left(1 + 2^{-2^n}\right)$$

$$= \lim_{k \to \infty} \sum_{m=0}^{2^k - 1} 2^{-m}$$

$$= \sum_{m=0}^{\infty} 2^{-m}$$

$$= 2.$$

The second equality can be verified by induction.

4.7 Clearly, the formula holds for $n = 0$ and $n = 1$. Assume that it holds
for n. Then

$$(x + y)^{(n+1)} = (x + y)^{(n)}[(x + y) - n]$$

$$= \sum_{k=0}^{n} \binom{n}{k} x^{(k)} y^{(n-k)} [(x - k) + y - (n - k)]$$

$$= \sum_{k=0}^{n} \binom{n}{k} x^{(k+1)} y^{(n-k)} + \sum_{k=0}^{n} \binom{n}{k} y^{(n-k+1)} x^{(k)}$$

$$= \sum_{k=1}^{n+1} \binom{n}{k-1} x^{(k)} y^{(n-k+1)} + \sum_{k=0}^{n} \binom{n}{k} y^{(n-k+1)} x^{(k)}$$

$$= \sum_{k=1}^{n} \left[\binom{n}{k-1} + \binom{n}{k} \right] x^{(k)} y^{(n+1-k)}$$

$$+ x^{(n+1)} y^{(0)} + y^{(n+1)} x^{(0)}$$

$$= \sum_{k=0}^{n+1} \binom{n+1}{k} x^{(k)} y^{(n+1-k)}.$$

Thus the formula holds for $n + 1$ and hence for all n by induction.

4.8 We prove the stronger result that $\sum_{k=1}^{n-1} \frac{1}{k} > \ln n$ for $n \geq 2$. The assertion is true for $n = 2$: $1 > \ln 2$. Assume that the result holds for n. Now $\sum_{k=1}^{n} \frac{1}{k} > \ln n + \frac{1}{n}$. We must show that $\ln n + \frac{1}{n} > \ln(n + 1)$. But this is equivalent to the inequality $e^{1/n} > 1 + 1/n$, which is trivially true.

Note: We can also solve the problem by comparing $\sum_{k=1}^{n-1} \frac{1}{k}$ and $\int_{1}^{n} \frac{1}{x} \, dx$.

4.9 Since $1 < \ln 3$, it suffices to prove that $\frac{1}{n} < \ln\left(\frac{2n+1}{2n-1}\right)$ for every positive integer n.

Solution (1):

$$\ln\left(\frac{2n+1}{2n-1}\right) = \ln\left(\frac{n + \frac{1}{2}}{n - \frac{1}{2}}\right)$$

$$= \int_{n-\frac{1}{2}}^{n+\frac{1}{2}} \frac{du}{u}$$

$$= \int_{0}^{\frac{1}{2}} \left(\frac{1}{n+t} + \frac{1}{n-t}\right) dt$$

$$= \int_{0}^{\frac{1}{2}} \frac{2n}{n^2 - t^2} \, dt$$

$$> \int_0^{\frac{1}{2}} \frac{2n}{n^2} \, dt$$

$$= \frac{2}{n} \int_0^{\frac{1}{2}} dt$$

$$= \frac{1}{n}.$$

Solution (2): Since for $|x| < 1$ we have

$$\ln\left(\frac{1+x}{1-x}\right) = \ln(1+x) - \ln(1-x)$$

$$= 2\left\{ x + \frac{x^3}{3} + \frac{x^5}{5} + \frac{x^7}{7} + \cdots \right\},$$

it follows that

$$\ln\left(\frac{2n+1}{2n-1}\right) = \ln\left(\frac{1+\frac{1}{2n}}{1-\frac{1}{2n}}\right)$$

$$= 2\left\{ \frac{1}{2n} + \frac{1}{3}\frac{1}{(2n)^3} + \frac{1}{5}\frac{1}{(2n)^5} + \frac{1}{7}\frac{1}{(2n)^7} + \cdots \right\}$$

$$> \frac{1}{n}.$$

4.10 The proof is by induction on n and makes use of the well-known explicit formula $f_n = \frac{1}{\sqrt{5}}(a^n - b^n)$ where $a = \frac{1}{2}(1 + \sqrt{5})$ and $b = \frac{1}{2}(1 - \sqrt{5})$. (See Example 13.3.) Note that $a \cdot b = -1$.

For $n = 1$, $\frac{1}{10}f_5 + \frac{3}{5}f_0 + \frac{1}{2} = \frac{5}{10} + 0 + \frac{1}{2} = 1$, as required.

Now assume that $\displaystyle\sum_{i=1}^{k} f_i^3 = \frac{1}{10}f_{3k+2} + \frac{3}{5}(-1)^{k-1}f_{k-1} + \frac{1}{2}$ for some $k \geq 1$. Then

$$f_{k+1}^3 = \frac{1}{5\sqrt{5}}(a^{k+1} - b^{k+1})^3$$

$$= \frac{1}{5\sqrt{5}}(a^{3k+3} - 3a^{2k+2}b^{k+1} + 3a^{k+1}b^{2k+2} - b^{3k+3})$$

$$= \frac{1}{5\sqrt{5}}(a^{3k+3} - b^{3k+3}) - \frac{1}{5\sqrt{5}}(3a^{k+1}b^{k+1})(a^{k+1} - b^{k+1})$$

$$= \frac{1}{5}f_{3k+3} - \frac{3}{5}(-1)^{k+1}f_{k+1},$$

and therefore

$$\sum_{i=1}^{k+1} f_i^3 = \sum_{i=1}^{k} f_i^3 + f_{k+1}^3$$

$$= \frac{1}{10}f_{3k+2} + \frac{3}{5}(-1)^{k-1}f_{k-1} + \frac{1}{2} + \frac{1}{5}f_{3k+3} - \frac{3}{5}(-1)^{k+1}f_{k+1}$$

$$= \frac{1}{10}f_{3k+2} + \frac{1}{10}f_{3k+3} + \frac{1}{10}f_{3k+3} + \frac{3}{5}(-1)^k(f_{k+1} - f_{k-1}) + \frac{1}{2}$$

$$= \frac{1}{10}f_{3k+4} + \frac{1}{10}f_{3k+3} + \frac{3}{5}(-1)^k f_k + \frac{1}{2}$$

$$= \frac{1}{10}f_{3k+5} + \frac{3}{5}(-1)^k f_k + \frac{1}{2}.$$

4.11 (a) We use induction on m. For $m = 1$,

$$f_m f_{n+1} + f_{m-1} f_n = 1 \cdot f_{n+1} + 0 \cdot f_n = f_{n+1}$$

and for $m = 2$,

$$f_m f_{n+1} + f_{m-1} f_n = 1 \cdot f_{n+1} + 1 \cdot f_n = f_{n+2}.$$

For the induction step,

$$f_{n+m+2} = f_{n+m} + f_{n+m+1}$$
$$= f_m f_{n+1} + f_{m-1} f_n + f_{m+1} f_{n+1} + f_m f_n$$
$$= (f_m + f_{m+1})f_{n+1} + (f_{m-1} + f_m)f_n$$
$$= f_{m+2} f_{n+1} + f_{m+1} f_n.$$

(b) We show by induction (on m) that $f_a | f_{ma}$. The result is clear for $m = 1$. For $m = 2$, by (a),

$$f_{2a} = f_{a+a} = f_a f_{a+1} + f_{a-1} f_a$$

and hence $f_a | f_{2a}$.

Assuming that $f_a | f_{ma}$, by (a) we have

$$f_{(m+1)a} = f_{ma+a} = f_{ma} f_{a+1} + f_{ma-1} f_a,$$

and therefore $f_a | f_{(m+1)a}$.

(c) We first consider the case where a and b are consecutive, say $b = a + 1$. In this case,

$$f_{\gcd(a,b)} = f_1 = 1.$$

Also, $\gcd(f_a, f_{a+1}) = 1$, since if d is a common divisor of f_a and f_{a+1} then $d | f_{a-1} = f_{a+1} - f_a$ and then by iteration we obtain $d | f_1 = 1$.

For the general case, let $g = \gcd(f_a, f_b)$ and $d = \gcd(a, b)$. We will show that $g | f_d$ and $f_d | g$. It will then follow that $g = f_d$.

By (b), we have $f_d | f_a$ and $f_d | f_b$ and hence $f_d | g$.

Integers x and y (of different signs) exist so that $ax + by = d$. We argue the case $x > 0$ and $y < 0$, the alternative case being similar.

Using (a) and (b), we have

$$f_{ax} = f_{d+b(-y)} = f_d f_{b(-y)+1} + f_{d-1} f_{b(-y)}.$$

Now $g | f_{ax}$ and $g | f_{b(-y)}$, and hence $g | f_d f_{b(-y)+1}$. But $\gcd(g, f_{b(-y)+1}) = 1$ since $g | f_{b(-y)}$ and $\gcd(f_{b(-y)}, f_{b(-y)+1}) = 1$. Therefore $g | f_d$.

4.12 For $n = 1$ and $j = 0$, $(D - a)e^{ax} = ae^{ax} - ae^{ax} = 0$.

Assuming that the result is true for some n, then for $0 \leq j \leq n$,

$$
\begin{aligned}
(D - a)^{n+1}(x^j e^{ax}) &= (D - a)^n (D - a)(x^j e^{ax}) \\
&= (D - a)^n (ax^j e^{ax} + jx^{j-1} e^{ax} - ax^j e^{ax}) \\
&= (D - a)^n (jx^{j-1} e^{ax}) \\
&= j(D - a)^n (x^{j-1} e^{ax}) \\
&= j \cdot 0 \\
&= 0.
\end{aligned}
$$

4.13 Solution (1): Note that the result is immediate if the r_j values are not distinct since in that case the product is zero, and also the determinant is zero since it has identical columns. For $n = 2$, $V_2 = r_2 - r_1$, so the result is true for $n = 2$.

Assume that the result holds for n real numbers and consider $n + 1$ distinct real numbers $r_1, \ldots, r_n, r_{n+1}$. Define

$$
f(x) = \begin{vmatrix}
1 & 1 & \cdots & 1 & 1 \\
r_1 & r_2 & \cdots & r_n & x \\
r_1^2 & r_2^2 & \cdots & r_n^2 & x^2 \\
\vdots & \vdots & \vdots & \vdots & \vdots \\
r_1^n & r_2^n & \cdots & r_n^n & x^n
\end{vmatrix}.
$$

Then f is a polynomial of degree n with zeros $r_1, r_2, \ldots, r_n$ and leading coefficient V_n. Hence $f(x) = V_n(x - r_1)(x - r_2)\ldots(x - r_n)$ and

$$
\begin{aligned}
V_{n+1} &= f(r_{n+1}) \\
&= \prod_{1 \le i < j \le n} (r_j - r_i) \\
&\quad \cdot (r_{n+1} - r_1) \cdot (r_{n+1} - r_2) \cdot \cdots \cdot (r_{n+1} - r_n) \\
&= \prod_{1 \le i < j \le n+1} (r_j - r_i).
\end{aligned}
$$

Solution (2): Again, we use induction on n, having already proved the base case above. Suppose that $V_n = \prod_{1 \le i < j \le n}(r_j - r_i)$ and consider

$$
V_{n+1} = \begin{vmatrix}
1 & 1 & \cdots & 1 \\
r_1 & r_2 & \cdots & r_{n+1} \\
r_1^2 & r_2^2 & \cdots & r_{n+1}^2 \\
\vdots & \vdots & \vdots & \vdots \\
r_1^{n-1} & r_2^{n-1} & \cdots & r_{n+1}^{n-1} \\
r_1^n & r_2^n & \cdots & r_{n+1}^n
\end{vmatrix}.
$$

Subtract r_{n+1} times the nth row from the $(n+1)$st row of this determinant. Then

$$
V_{n+1} = \begin{vmatrix}
1 & \cdots & 1 & 1 \\
r_1 & \cdots & r_n & r_{n+1} \\
r_1^2 & \cdots & r_n^2 & r_{n+1}^2 \\
\vdots & \vdots & \vdots & \vdots \\
r_1^{n-1} & \cdots & r_n^{n-1} & r_{n+1}^{n-1} \\
r_1^n - r_{n+1}r_1^{n-1} & \cdots & r_n^n - r_{n+1}r_n^{n-1} & 0
\end{vmatrix}.
$$

Now subtract r_{n+1} times the $(n-1)$st row from the nth row to obtain

$$
V_{n+1} = \begin{vmatrix}
1 & \cdots & 1 & 1 \\
r_1 & \cdots & r_n & r_{n+1} \\
r_1^2 & \cdots & r_n^2 & r_{n+1}^2 \\
\vdots & \vdots & \vdots & \vdots \\
r_1^{n-1} - r_{n+1}r_1^{n-2} & \cdots & r_n^{n-1} - r_{n+1}r_n^{n-2} & 0 \\
r_1^n - r_{n+1}r_1^{n-1} & \cdots & r_n^n - r_{n+1}r_n^{n-1} & 0
\end{vmatrix}.
$$

Continue in this way to obtain

$$
V_{n+1} = \begin{vmatrix}
1 & \cdots & 1 & 1 \\
r_1 - r_{n+1} & \cdots & r_n - r_{n+1} & 0 \\
r_1^2 - r_{n+1}r_1 & \cdots & r_n^2 - r_{n+1}r_n & 0 \\
\vdots & \vdots & \vdots & \vdots \\
r_1^{n-1} - r_{n+1}r_1^{n-2} & \cdots & r_n^{n-1} - r_{n+1}r_n^{n-2} & 0 \\
r_1^n - r_{n+1}r_1^{n-1} & \cdots & r_n^n - r_{n+1}r_n^{n-1} & 0
\end{vmatrix} .
$$

Now expand by the last column, factor terms out of columns, and apply the induction hypothesis:

$$
\begin{aligned}
V_{n+1} &= \prod_{i=1}^{n}(r_i - r_{n+1})(-1)^n \begin{vmatrix}
1 & 1 & \cdots & 1 \\
r_1 & r_2 & \cdots & r_n \\
\vdots & \vdots & \vdots & \vdots \\
r_1^{n-2} & r_2^{n-2} & \cdots & r_n^{n-2} \\
r_1^{n-1} & r_2^{n-1} & \cdots & r_n^{n-1}
\end{vmatrix} \\
&= \prod_{i=1}^{n}(r_{n+1} - r_i) \prod_{1 \le i < j \le n}(r_j - r_i) \\
&= \prod_{1 \le i < j \le n+1}(r_j - r_i).
\end{aligned}
$$

Note: The Vandermonde determinant formula holds for $r_1, \ldots, r_n$ elements of any field.

4.14 We prove the result by induction on the size of the number's numerator. All fractions $1/s$ are certainly representable in the desired way. Assume that all fractions $r/s < 1$ are representable if $r < R$. Consider $R/s < 1$. Define m by $1/m \le R/s < 1/(m-1)$. Since $R(m-1) < s$, it follows that $Rm - s < R$, and hence $(Rm - s)/sm = R/s - 1/m$ is representable. Furthermore, since

$$
\frac{R}{s} - \frac{1}{m} < \frac{1}{m-1} - \frac{1}{m} = \frac{1}{m(m-1)} \le \frac{1}{m},
$$

the representation of $Rm - s$ does not use $1/m$. Hence R/s is representable, and, by induction, all fractions $r/s < 1$ can be represented as required. For rational numbers r/s greater than or equal to 1, add as many terms as possible of the harmonic series $1 + \frac{1}{2} + \frac{1}{3} + \frac{1}{4} + \cdots$ until the first term $1/m$ will not fit. Then apply the algorithm given above to

$$
\frac{r}{s} - 1 - \frac{1}{2} - \frac{1}{3} - \cdots - \frac{1}{m-1},
$$

noting that none of the fractions $1/q$, with $q \leq m$, will be chosen.

Note: The proof above gives a practical algorithm for finding representations of rational numbers r/s as sums of Egyptian fractions. Just subtract the largest possible fraction $1/n$ from the given rational number r/s and continue the process with the smaller rational number $r/s - 1/n$. This procedure is an example of a "greedy algorithm" because the problem is reduced as much as possible at each step. Some experimentation with this algorithm shows that it may produce representations with very large denominators. However, other algorithms exist which may produce smaller denominators. One such possibility is to first write

$$\frac{p}{q} = \frac{1}{q} + \cdots + \frac{1}{q},$$

where there are p terms in the summation. Leaving the first term unaltered, we write the second term as

$$\frac{1}{q+1} + \frac{1}{q(q+1)}.$$

Neither of the summands has been previously used. Expanding on this representation, we write the next term as

$$\frac{1}{q+2} + \frac{1}{(q+2)(q+3)} + \frac{1}{q^2+q+1} + \frac{1}{(q^2+q)(q^2+q+1)}.$$

Continuing this process we eventually arrive at the desired representation, although the process may have to be iterated extra times in order to avoid repetitions of fractions. (The proof that the process terminates is quite difficult.)

Additional Problems

4.15 Show that $2^n \geq 1 + n$ for all $n \geq 1$.

4.16 For the Fibonacci numbers f_n, show that

$$f_1 + f_2 + \cdots + f_n = f_{n+2} - 1.$$

4.17 Given n lines in the plane, show that the regions into which the plane is divided can be colored with two colors so that neighboring regions (regions sharing a line segment boundary) are different colors.

4.18 Solve the previous problem with 'line' replaced by 'circle'.

Note: This problem and the previous one can be solved using parity arguments. See Additional Problems 7.16 and 7.17.

4.19 Suppose that n -1's and n $+1$'s are distributed around a circle. Show that it is always possible to start at one of the numbers and go around the circle in such a way that the partial sums of the numbers passed are all nonnegative.

4.20 (Bernoulli's inequality) Suppose that x is a real number greater than -1 and n is a positive integer. Show that

$$(1 + x)^n \geq 1 + nx.$$

4.21 Prove that every positive integer greater than 6 is the sum of one or more distinct prime numbers. For example,

$$7 = 7, \ 8 = 3 + 5, \ 9 = 2 + 7, \ 10 = 3 + 7, \ 11 = 11, \ 12 = 5 + 7.$$

(Assume Bertrand's postulate: for every $n \geq 2$, there exists a prime number between n and $2n$.)

4.22 For each $k \geq 1$, define

$$S_k(n) = \sum_{i=1}^{n} i^k.$$

(a) Prove the following formulas:

$$S_1(n) \quad = \quad \frac{n(n+1)}{2},$$

$$S_2(n) \quad = \quad \frac{n(n+1)(2n+1)}{6},$$

$$S_3(n) \quad = \quad \left[\frac{n(n+1)}{2}\right]^2.$$

(b) Show that $S_k(n)$ is a polynomial in n of degree $k+1$ and leading coefficient $1/(k+1)$.

4.23 (a) Given a simple (non-self-intersecting) polygon in the plane, and a point P in the interior of the polygon, show that some vertex v of the polygon is "visible" from P; that is, the line segment joining P and v touches the polygon only at v.

(b) Show that there do not necessarily exist two consecutive vertices of the polygon, v and w, which are both visible from P.

4.24 (U. S. A. Olympiad, 1978; modified) An integer n is called *good* if we can write $n = a_1 + a_2 + \cdots + a_k$ where a_1, a_2, ..., a_k are positive integers (not necessarily distinct) satisfying

$$\frac{1}{a_1} + \frac{1}{a_2} + \cdots + \frac{1}{a_k} = 1.$$

Given that it is known that the integers 33 through 73 are good, prove that every positive integer greater than or equal to 33 is good.

4.25 Prove that for each $n \geq 2$, there exists a permutation $(p_1, p_2, \ldots, p_n)$ of $(1, 2, \ldots, n)$ such that $p_{k+1} \mid (p_1 + p_2 + \cdots + p_k)$ for $1 \leq k \leq n - 1$.

4.26 Let $f(x) = (x^2 - 1)^{1/2}$, for $x > 1$. Show that $f^{(n)}(x) > 0$ for odd n and $f^{(n)}(x) < 0$ for even n.

4.27 Suppose that $\{x_n\}$ is a sequence of positive real numbers for which

$$\sum_{i=1}^{n} x_i^3 = \left(\sum_{i=1}^{n} x_i \right)^2$$

for each $n \geq 1$. Prove that $x_n = n$ for all $n \geq 1$.

4.28 Show that

$$\sum_{k=1}^{n} \cos kx = \frac{\cos[(n + 1)x/2] \sin(nx/2)}{\sin(x/2)},$$

if $\sin(x/2) \neq 0$.

4.29 Let $a_1 = 1$ and $a_{n+1} = \sqrt{a_1 + a_2 + \cdots + a_n}$ for $n \geq 1$. Prove that

$$\lim_{n \to \infty} \frac{a_n}{n} = \frac{1}{2}.$$

Hint: Use induction to show that

$$\frac{n - \sqrt{n}}{2} \leq a_n \leq \frac{n}{2}$$

for $n > 1$.

4.30 (*Math. Magazine*, Problem 1361, December 1990). Does there exist a differentiable function f, defined for all real numbers x, such that $f(f(x)) = e^x$ for all x? If so, exhibit such a function; if not, show why not.

4.31 (*Math. Magazine*, Problem 1146, May 1982; modified). Let $f(x) = 2^x$ and $g(x) = 3^x$ for all real x, and indicate iteration by superscripts. It is easy to check that

$$f^3(1) < g^2(1) < f^4(1) < g^3(1) < f^5(1) < g^4(1).$$

Is is true that $f^n(1) < g^{n-1}(1) < f^{n+1}(1)$ for all $n \geq 3$?

4.32 Suppose that an $a \times b$ (with at least one of a or b even) rectangle has been tiled with 1×2 dominoes. Show that in this tiling there exists a 2×2 square composed of two horizontal dominoes or two vertical dominoes.

Chapter 5

Specialization and Generalization

"The art of detection is finding a common denominator for the fractions of a case."

ELSA BARKER
Dexter Drake, *The C. I. D. of Dexter Drake*, 1929

Sometimes it is best to solve a special case of a problem before trying to solve the problem in its entirety.

Example 5.1. Show that a pair of dice may not be weighted so as to give all sums $2, \ldots, 12$ with equal probability.

Solution: The natural way to set up the problem is to assume that die P takes the values 1, 2, 3, 4, 5, 6 with probabilities p_1, p_2, p_3, p_4, p_5, p_6, respectively, and die Q takes the values 1, 2, 3, 4, 5, 6 with probabilities q_1, q_2, q_3, q_4, q_4, q_5, q_6, respectively. However, it may not be clear what to do next, so we specialize by assuming that $p_i = q_i$ for $1 \leq i \leq 6$. If all sums occur with equal probability, then

$$p_1^2 = \tfrac{1}{11} \tag{5.1}$$

$$p_1 p_2 + p_2 p_1 = \tfrac{1}{11} \tag{5.2}$$

$$p_1 p_3 + p_2 p_2 + p_3 p_1 = \tfrac{1}{11} \tag{5.3}$$

$$p_1 p_4 + p_2 p_3 + p_3 p_2 + p_4 p_1 = \tfrac{1}{11} \tag{5.4}$$

$$p_1 p_5 + p_2 p_4 + p_3 p_3 + p_4 p_2 + p_5 p_1 = \tfrac{1}{11} \tag{5.5}$$

$$p_1p_6 + p_2p_5 + p_3p_4 + p_4p_3 + p_5p_2 + p_6p_1 = \tfrac{1}{11} \qquad (5.6)$$

$$p_2p_6 + p_3p_5 + p_4p_4 + p_5p_3 + p_6p_2 = \tfrac{1}{11} \qquad (5.7)$$

$$p_3p_6 + p_4p_5 + p_5p_4 + p_6p_3 = \tfrac{1}{11} \qquad (5.8)$$

$$p_4p_6 + p_5p_5 + p_6p_4 = \tfrac{1}{11} \qquad (5.9)$$

$$p_5p_6 + p_6p_5 = \tfrac{1}{11} \qquad (5.10)$$

$$p_6^2 = \tfrac{1}{11} \qquad (5.11)$$

The equations (5.1) and (5.11) imply that $p_1 = 1/\sqrt{11}$ and $p_6 = 1/\sqrt{11}$. However, these results are incompatible with the equation (5.6). Hence all sums do not occur with equal probability.

Now we return the original problem. Again, eleven equations hold:

$$p_1q_1 = \tfrac{1}{11} \qquad (5.12)$$

$$p_1q_2 + p_2q_1 = \tfrac{1}{11} \qquad (5.13)$$

$$p_1q_3 + p_2q_2 + p_3q_1 = \tfrac{1}{11} \qquad (5.14)$$

$$p_1q_4 + p_2q_3 + p_3q_2 + p_4q_1 = \tfrac{1}{11} \qquad (5.15)$$

$$p_1q_5 + p_2q_4 + p_3q_3 + p_4q_2 + p_5q_1 = \tfrac{1}{11} \qquad (5.16)$$

$$p_1q_6 + p_2q_5 + p_3q_4 + p_4q_3 + p_5q_2 + p_6q_1 = \tfrac{1}{11} \qquad (5.17)$$

$$p_2q_6 + p_3q_5 + p_4q_4 + p_5q_3 + p_6q_2 = \tfrac{1}{11} \qquad (5.18)$$

$$p_3q_6 + p_4q_5 + p_5q_4 + p_6q_3 = \tfrac{1}{11} \qquad (5.19)$$

$$p_4q_6 + p_5q_5 + p_6q_4 = \tfrac{1}{11} \qquad (5.20)$$

$$p_5q_6 + p_6q_5 = \tfrac{1}{11} \qquad (5.21)$$

$$p_6q_6 = \tfrac{1}{11}. \qquad (5.22)$$

As in the solution of the special case, we look for a contradiction among these equations. In the special case, we found that the equations (5.1), (5.6), and (5.11) are incompatable. Perhaps the corresponding equations in the general case, (5.12), (5.17), and (5.22), are incompatable also. From (5.12) and (5.22), $p_1q_1 = 1/11$ and $p_6q_6 = 1/11$. In order to obtain a contradiction with (5.17), it looks as though we must relate the quantities p_1q_1, p_6q_6, and $p_1q_6 + p_6q_1$. This we can do via the arithmetic mean–geometric mean inequality (Chapter 15). Thus,

$$\frac{p_1q_6 + p_6q_1}{2} \geq \sqrt{p_1q_6p_6q_1} = \sqrt{p_1q_1p_6q_6} = \frac{1}{11},$$

i.e.,

$$p_1 q_6 + q_1 p_6 \geq \frac{2}{11},$$

contradicting (5.17). Again, not all sums occur with the same probability.

Note: The above argument can be generalized to show that two n-sided dice cannot be weighted so as to give all sums $2, \ldots, 2n$ with equal probability. Further generalizations are possible. ∎

Example 5.2. Given a point P on the ellipse

$$\frac{x^2}{a^2} + \frac{y^2}{b^2} = 1,$$

show how to find the two points Q and R on the ellipse such that the triangle PQR has maximum possible area.

Solution: The solution would be simple if the point P were on a circle instead of an ellipse, for an equilateral triangle has the maximum possible area of any triangle inscribed in a circle. Therefore, let us make a change of variables that transforms the ellipse into a circle:

$$x' = \frac{b}{a}x, \quad y' = y.$$

The new equation is

$$x'^2 + y'^2 = b^2,$$

i.e., the equation for a circle with radius b. Since the change of variables multiplies the area of any plane figure by b/a, the triangle of maximum area for the ellipse corresponds to the triangle of maximum area for the circle. The solution for the circle is the equilateral triangle PQR and the solution for the ellipse is the inverse transformation of this triangle. ∎

Sometimes it is easiest to solve a problem by first generalizing it. Look closely at the hypothesis and conclusion of the result to be proved. Can either (or both) be generalized?

Example 5.3. Show that some multiple of the integer 17623176 involves all ten digits.

Solution: The number 17623176 is a red herring. The result is true for *any* positive integer n. For suppose that $n < 10^k$. Then some multiple of n lies between $1234567890 \cdot 10^k$ and $1234567891 \cdot 10^k$, and hence contains all ten digits. ∎

Example 5.4. Prove that there are integers p_n, q_n for which

$$\int_0^3 e^x \frac{x^n(3-x)^n}{n!}\,dx = p_n - q_n e^3$$

for all $n \geq 0$.

Solution: We have

$$I_0 = \int_0^3 e^x\,dx = e^3 - 1,$$

which is expressible in the desired form with $p_0 = q_0 = -1$. Now

$$\begin{aligned}
\int_0^3 e^x \frac{x^n(3-x)^n}{n!}\,dx &= \left. \frac{1}{n!}e^x(3x - x^2)^n \right]_0^3 \\
&\quad - \frac{1}{n!}\int_0^3 e^x n(3x - x^2)^{n-1}(3 - 2x)\,dx \\
&= -\frac{1}{(n-1)!}\int_0^3 e^x(3x - x^2)^{n-1}(3 - 2x)\,dx \\
&= 3I_{n-1} + \frac{2}{(n-1)!}\int_0^3 e^x x(3x - x^2)^{n-1}\,dx.
\end{aligned}$$

We are not making progress. Something more is needed.

For nonnegative integers a and b, let us define a new function

$$f(a,b) = \int_0^3 \frac{e^x x^a(3-x)^b}{a!b!}\,dx,$$

and show that there exist integers p and q (depending on a and b) such that $M!f(a,b) = p + qe^3$ where $M = \max\{a,b\}$. The result then follows since

$$\int_0^3 \frac{e^x x^n(3-x)^n}{n!}\,dx = n!f(n,n).$$

The proof is by "double induction" on a and b. If $a = b = 0$, then

$$0!f(0,0) = \int_0^3 e^x\,dx = -1 + e^3.$$

Suppose that $k!f(0,k) = p + qe^3$ for some integers p and q. Using integration by parts,

$$f(0, k+1) = \frac{3^{k+1}}{(k+1)!} + f(0,k) = \frac{3^{k+1}}{(k+1)!} + \frac{1}{k!}(p + qe^3),$$

so

$$(k+1)!f(0,k+1) = [3^{k+1} + (k+1)p] + (k+1)qe^3.$$

Similarly, if $k!f(k,0)$ has the form $p + qe^3$, then

$$f(k+1,0) = \frac{3^{k+1}e^3}{(k+1)!} - f(k,0) = \frac{3^{k+1}e^3}{(k+1)!} - \frac{1}{k!}(p+qe^3),$$

so

$$(k+1)!f(k+1,0) = -(k+1)p + [3^{k+1} - (k+1)q]e^3.$$

Now we assume that the result holds for all i and j with $0 \le i \le a$, $0 \le j \le b$, and either $i < a$ or $j < b$. Then (again using integration by parts) it follows that

$$f(a,b) = f(a,b-1) - f(a-1,b).$$

By the induction assumption, there exist integers p_1, q_1, p_2, q_2 such that

$$f(a,b-1) = \frac{1}{\alpha!}\left(p_1 + q_1 e^3\right) \quad \text{and} \quad f(a-1,b) = \frac{1}{\beta!}\left(p_2 + q_2 e^3\right)$$

where $\alpha = \max\{a, b-1\}$ and $\beta = \max\{a-1, b\}$. If $M = \max\{a,b\}$, then $M \ge \alpha$ and $M \ge \beta$ and hence $M!/\alpha!$ and $M!/\beta!$ are integers, and therefore

$$M!f(a,b) = \frac{M!}{\alpha!}\left(p_1 + q_1 e^3\right) - \frac{M!}{\beta!}\left(p_2 + q_2 e^3\right)$$

has the desired form. ∎

Problems

5.1 (Putnam Competition, 1982) Evaluate

$$\int_0^\infty \frac{\text{Arctan}(\pi x) - \text{Arctan } x}{x}\, dx.$$

5.2 Let $\pi(n)$ be the number of primes not greater than n. Prove that $\pi(n)$ divides n infinitely often.

5.3 (Putnam Competition, 1990) Is $\sqrt{2}$ the limit of a sequence of numbers of the form $n^{1/3} - m^{1/3}$ $(n,m = 0,1,2,\ldots)$? Justify your answer.

5.4 Suppose that $x_1 = 0$ and

$$x_n = x_{\lfloor n/2 \rfloor} + x_{\lfloor (n+1)/2 \rfloor} + 2$$

for $n > 1$. Find an explicit formula for x_n.

5.5 Find

$$\lim_{n\to\infty}\left(\int_0^2 e^{nx^2}\,dx\right)^{1/n}.$$

5.6 Show that $f(x) = \sin x$ is uniformly continuous (see Glossary) on $\mathbf{R}$.

5.7 Let $H = \{1, \frac{1}{2}, \frac{1}{3}, \frac{1}{4}, \ldots\}$ and define $g\colon \mathbf{R} \to \mathbf{R}$ where

$$g(x) = \inf\{|x - t|\colon t \in H\}.$$

Show that g is uniformly continuous on $\mathbf{R}$.

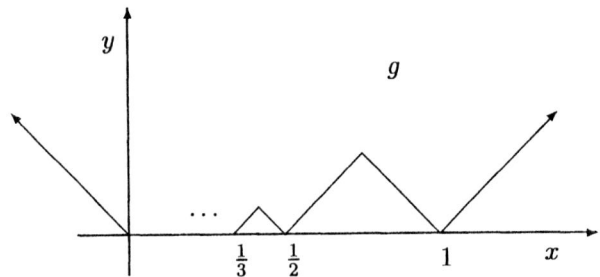

Solutions

5.1 For $a > 0$, define

$$f(a) = \int_0^\infty \frac{\text{Arctan } ax - \text{Arctan } x}{x}\,dx.$$

Then

$$f'(a) = \int_0^\infty \frac{1}{x} \cdot \frac{1}{1 + a^2 x^2} \cdot x\,dx = \frac{1}{a}\text{Arctan } ax\Big|_0^\infty = \frac{\pi}{2} \cdot \frac{1}{a}.$$

Hence $f(a) = \frac{\pi}{2}\ln a + C$. Letting $a = 1$, we find that $C = 0$. Thus $f(\pi) = \frac{\pi}{2}\ln \pi$.

5.2 The result holds for any integer-valued function f that satisfies the following three conditions:

(1) $f(2) > 0$;

(2) $f(n + 1) - f(n) = 0$ or 1 for each n;

(3) $n/f(n)$ tends to ∞.

For suppose that f satisfies these conditions. Let $g(n) = n - kf(n)$, where $k > 2$. From (1), $g(2)$ is negative, while from (3), $g(n)$ is positive for sufficiently large n. It follows from (2) that $g(n) = 0$ for some n. For such an n, $kf(n) = n$, so that n is divisible by $f(n)$. As k is arbitrary, $f(n)$ divides n infinitely often.

It is obvious that $\pi(n)$ satisfies the conditions (1) and (2) above. That $\pi(n)$ satisfies the condition (3) can be proved from first principles or from the prime number theorem (Glossary).

5.3 There is nothing special about $\sqrt{2}$ in this problem. We will show that every real number t is the limit of such a sequence. Set $n - m = 1$. Then

$$n^{1/3} - m^{1/3} \;=\; \frac{n - m}{n^{2/3} + n^{1/3}m^{1/3} + m^{2/3}}$$

$$\leq \;\frac{1}{3m^{1/3}}.$$

Let ϵ be an arbitrary positive number, and choose m so that $n^{1/3} - m^{1/3} < 2\epsilon$. Because the interval $(t - \epsilon, t + \epsilon)$ has length 2ϵ, there exists an integer k such that

$$t - \epsilon < k(n^{1/3} - m^{1/3}) = (nk^3)^{1/3} - (mk^3)^{1/3} < t + \epsilon.$$

We have proved the assertion that every real number is the limit of a sequence of the form $n^{1/3} - m^{1/3}$.

5.4 We first consider powers of 2. Let $y_j = x_{2^j}$ for $j \geq 0$. Then $y_0 = 0$ and for $j > 0$,

$$y_j = x_{2^j} = x_{2^{j-1}} + x_{2^{j-1}} + 2 = 2y_{j-1} + 2.$$

Iterating this result,

$$\begin{aligned}
y_j &= 2y_{j-1} + 2 \\
&= 2(2y_{j-2} + 2) + 2 \\
&= 2^2 y_{j-2} + 2^2 + 2 \\
&\;\;\vdots \\
&= 2^j y_0 + 2 + 2^2 + \cdots + 2^j \\
&= 2^{j+1} - 2 \\
&= 2 \cdot 2^j - 2
\end{aligned}$$

Thus if n is a power of 2, $x_n = 2n - 2$ and we now use induction to show this formula is valid for all n. The basis step is clear. Suppose that $x_j = 2j - 2$ for all $j < n$. If n is even, then

$$x_n = 2x_{n/2} + 2 = 2\left(2 \cdot \frac{n}{2} - 2\right) + 2 = 2n - 2.$$

If n is odd,

$$x_n = x_{(n-1)/2} + x_{(n+1)/2} + 2 = 2 \cdot \frac{n-1}{2} - 2 + 2 \cdot \frac{n+1}{2} - 2 + 2 = 2n - 2.$$

5.5 The answer is e^4. We will show, in general, that if f is any positive continuous function on $[a, b]$, then

$$\lim_{n \to \infty} \left(\int_a^b f(x)^n \, dx\right)^{1/n} = M,$$

where M is the maximum value of f on $[a, b]$.

Since $f(x) \le M$,

$$\left[\int_a^b f(x)^n \, dx\right]^{1/n} \le \left[\int_a^b M^n \, dx\right]^{1/n} = M \cdot (b-a)^{1/n},$$

and

$$\lim_{n \to \infty} M(b-a)^{1/n} = M.$$

Now suppose that $f(x_0) = M$ for some x_0 in (a, b). For any $\epsilon > 0$, there exists $\delta > 0$ so that $|x - x_0| < \delta$ implies both that x is in $[a, b]$ and $|f(x) - M| < \epsilon$, and hence $f(x) > M - \epsilon$. Thus

$$\left[\int_a^b f(x)^n \, dx\right]^{1/n} \ge \left[\int_{x_0 - \delta}^{x_0 + \delta} f(x)^n \, dx\right]^{1/n} \ge (M - \epsilon) \cdot (2\delta)^{1/n},$$

and

$$\lim_{n \to \infty} (M - \epsilon)(2\delta)^{1/n} = M - \epsilon.$$

Since ϵ is arbitrary, the result follows. (Note that only a minor adjustment is needed to handle the case of $x_0 = a$ or b.)

Back to the original problem, the function $f(x) = e^{x^2}$ is increasing on $[0, 2]$, so

$$\lim_{n \to \infty} \left(\int_0^2 e^{nx^2} \, dx\right)^{1/n} = f(2) = e^4.$$

5.6 A function f is said to satisfy a *uniform Lipschitz condition* on $\mathbf{R}$ provided that

$$|f(x) - f(y)| \leq M \cdot |x - y|$$

for some constant $M > 0$ and all x and y in $\mathbf{R}$. All uniformly Lipschitz functions are uniformly continuous since for any $\epsilon > 0$ we can choose $\delta = \epsilon/M$. Then $|x - y| < \delta$ implies that

$$|f(x) - f(y)| \leq M \cdot |x - y| < M \cdot \delta = \epsilon.$$

Any function (like $\sin x$) which has a bounded derivative is uniformly Lipschitz (and hence uniformly continuous) for if we have $|f'(t)| \leq M$ for all t, then (by the mean value theorem) we have

$$|f(x) - f(y)| = |f'(c)| \cdot |x - y|$$

for some c between x and y and therefore

$$|f(x) - f(y)| \leq M \cdot |x - y|.$$

5.7 More generally, let A be any nonempty subset of $\mathbf{R}$ and consider $f_A : \mathbf{R} \to \mathbf{R}$ where $f_A(x) = \inf\{|x - t| : t \in A\}$. We show that f_A is uniformly continuous on $\mathbf{R}$ by showing f_A is uniformly Lipschitz (see previous solution).

Suppose that $x, y \in \mathbf{R}$ and $t \in A$. Then

$$f_A(x) \leq |x - t| \leq |x - y| + |y - t|,$$

and hence $f_A(x) - |x - y| \leq |y - t|$, which implies that $f_A(x) - |x - y| \leq f_A(y)$. Therefore, $f_A(x) - f_A(y) \leq |x - y|$. Similarly, by interchanging x and y, we obtain $f_A(y) - f_A(x) \leq |x - y|$. It follows that $|f_A(x) - f_A(y)| \leq |x - y|$.

Additional Problems

5.8 Find a factorization of the number 999973.

5.9 Prove that

$$x_1^2 + \cdots + x_n^2 \geq x_1 x_n + x_2 x_{n-1} + \cdots + x_n x_1,$$

where $x_1, \ldots, x_n$ are real numbers.

5.10 Solve the equation $\sin^{2000} x + \cos^{2000} x = 1$.

5.11 Let π be a permutation of the set $\{1, \ldots, 100\}$. The *order* of π is the smallest positive integer k such that π^k (π composed with itself k times) is the identity permutation. What is the greatest possible order of π?

5.12 (Putnam Competition, 1964) Let p_n ($n = 1, 2, \ldots$) be a bounded sequence of integers which satisfies the recursion

$$p_n = \frac{p_{n-1} + p_{n-2} + p_{n-3}p_{n-4}}{p_{n-1}p_{n-2} + p_{n-3} + p_{n-4}}.$$

Show that the sequence eventually becomes periodic.

Hint: Show that the result holds for *any* bounded sequence of integers $\{p_n\}$ satisfying *any* recurrence relation

$$p_n = f(p_{n-1}, \ldots, p_{n-k}).$$

5.13 Show that if every point of $\mathbf{R}^2$ is colored one of three colors, there exist two points of the same color one unit apart.

5.14 (Ramsey's theorem) Denote by K_n the graph with n vertices every pair of which is joined by an edge. We call K_n the *complete graph* on n vertices. Given $m \geq 2$, show that there exists an integer n with the property that every coloring of the edges of K_n with two colors yields a subgraph K_m all of whose edges are the same color.

Chapter 6

Symmetry

"She had a seventy-eight-inch bust, forty-six-inch waist, and seventy-two-inch hips—measurements that were exactly right, I thought, for her height of eleven feet, four inches."

RICHARD PRATHER
Shell Scott (viewing a statue),
Take a Murder, Darling, 1958

In this chapter we examine problems whose solutions use the general concept of symmetry.

Example 6.1. Two points A and B lie on one side of a line l. Show how to construct the shortest path from A to B which touches l. See Figure 6.1.

Solution: As in Figure 6.2(a), let B' be the symmetric point to B on the other side of l. The shortest path from A to B' is the straight line AB'. Suppose that this line intersects l at point P. We claim that APB is the

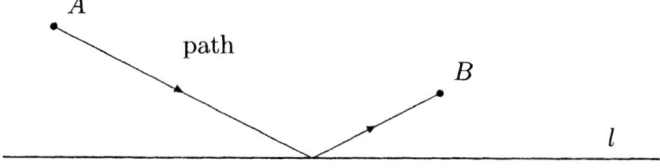

Figure 6.1: A path from A to B.

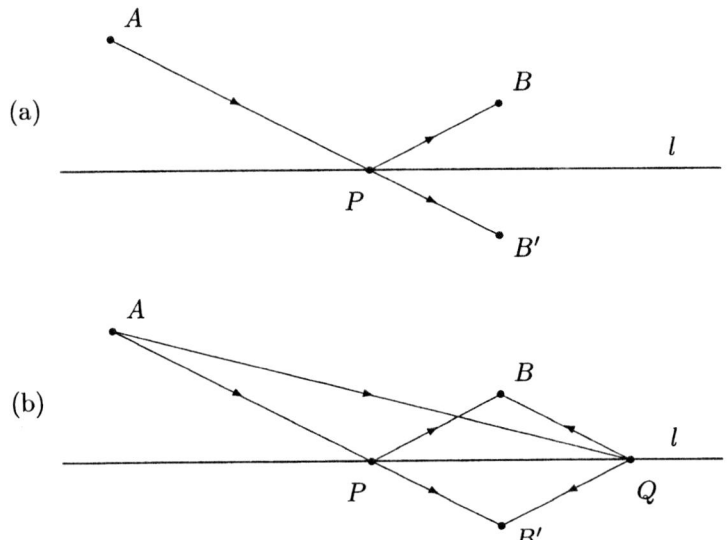

(a)

(b)

Figure 6.2: (a) A path via reflection; (b) the path APB and a longer path (AQB).

shortest path from A to B which touches l. If Q is any other any other point on l, the path AQB is longer than APB:

$$|AQB| = |AQB'| > |APB'| = |APB|;$$

see Figure 6.2(b). Thus APB is the shortest path from A to B which touches l.   ∎

Example 6.2. Let P be a point inside an equilateral triangle ABC, with $PA = 4$, $PB = 3$, and $PC = 5$. Find x, the side length of the triangle. See Figure 6.3.

Solution: As in Figure 6.4, rotate $\triangle ABC$ $\pi/3$ radians clockwise around B. Suppose that P and C are transformed to P' and C', respectively. Since $\triangle PP'C$ has sides of lengths 3, 4, and 5, $\angle PP'C$ is a right angle. Hence

$$m(\angle BP'C) = \frac{\pi}{2} + \frac{\pi}{3} = \frac{5\pi}{6}.$$

By the law of cosines,

$$x^2 = 3^2 + 4^2 - 2 \cdot 3 \cdot 4 \cdot \cos\frac{5\pi}{6}$$

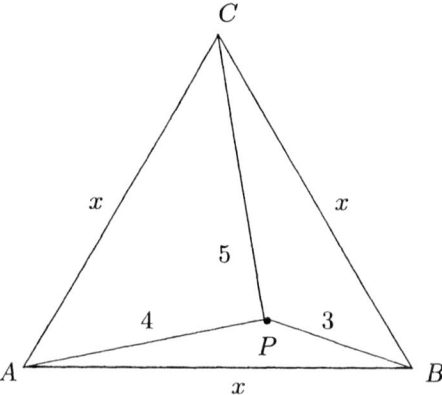

Figure 6.3: What is x?

$$= \ 25 + 12\sqrt{3},$$

and

$$x = \sqrt{25 + 12\sqrt{3}}.$$

■

Even problems not explicitly about geometry may have solutions which use symmetry.

Example 6.3. (The 50's game) Two players, Alpha and Beta, alternately choose positive integers between 1 and 6 (repetition allowed). The first player who makes the sum of all the chosen numbers equal 50 is the winner. Which player has a winning strategy? If 50 is replaced by any positive integer n, which player has a winner strategy?

Solution: The first player, Alpha, wins for all values of n not equal to 7. If $n < 7$, then Alpha chooses n and wins instantly. If $n = 7$, then Alpha chooses x and Beta chooses $7 - x$ and wins. If $n > 7$, then Alpha calculates the remainder when n is divided by 7 and chooses that number. Thereafter, whenever Beta chooses x, Alpha chooses $7 - x$. Thus the game is eventually reduced to the $n = 7$ game with Alpha going second; hence Alpha wins. ■

Example 6.4. (Putnam Competition, 1980) Evaluate

$$\int_0^{\pi/2} \frac{dx}{1 + (\tan x)^{\sqrt{2}}}.$$

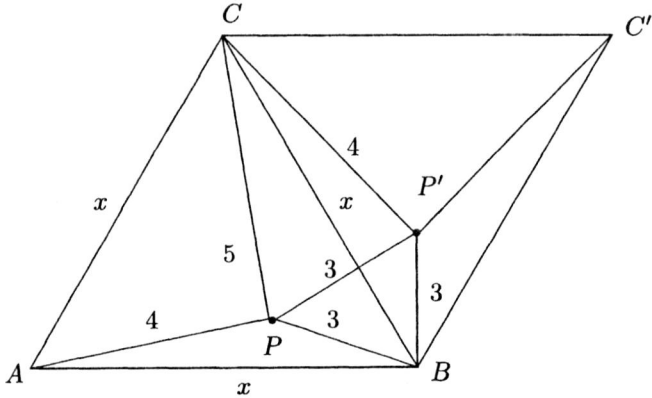

Figure 6.4: Use of a rotational symmetry.

Solution: Making the substitution $y = \pi/2 - x$, we obtain

$$\int_0^{\pi/2} \frac{dx}{1 + (\tan x)^{\sqrt{2}}} = \int_{\pi/2}^0 \frac{-dy}{1 + (\cot y)^{\sqrt{2}}}$$

$$= \int_0^{\pi/2} \frac{(\tan y)^{\sqrt{2}} \, dy}{(\tan y)^{\sqrt{2}} + 1}.$$

Therefore

$$2 \int_0^{\pi/2} \frac{dx}{1 + (\tan x)^{\sqrt{2}}} = \int_0^{\pi/2} \frac{dx}{1 + (\tan x)^{\sqrt{2}}} + \int_0^{\pi/2} \frac{(\tan x)^{\sqrt{2}} \, dx}{1 + (\tan x)^{\sqrt{2}}}$$

$$= \int_0^{\pi/2} \frac{1 + (\tan x)^{\sqrt{2}}}{1 + (\tan x)^{\sqrt{2}}} \, dx$$

$$= \int_0^{\pi/2} dx$$

$$= \frac{\pi}{2}$$

and hence

$$\int_0^{\pi/2} \frac{dx}{1 + (\tan x)^{\sqrt{2}}} = \frac{\pi}{4}.$$

Note: The exponent $\sqrt{2}$ is a red herring. Every positive exponent gives the same result. ∎

Problems

6.1 Rank Amateur challenges Grandmaster One and Grandmaster Two to a chess match. He stipulates that he will play one game against each of them and that he is to be considered the winner of the match if he scores at least one point. (In chess, a win earns one point, a draw one-half point, and a loss no points.) How does Rank Amateur win the match?

6.2 A function f is called *even* if $f(-x) = f(x)$ for all x; it is called *odd* if $f(-x) = -f(x)$ for all x. Show that every function f may be written as the sum of an even function and an odd function. For example,

$$e^x = \cosh x + \sinh x.$$

Furthermore, show that such a representation is always unique.

6.3 Consider a game in which Alpha flips $n + 1$ fair coins, Beta flips n fair coins, and the player who gets the most heads wins. Show that if Beta wins ties, then Alpha and Beta have equal chances of winning.

6.4 (U. S. A. Olympiad, 1975) A deck of n playing cards, which contains three aces, is shuffled at random (it is assumed that all possible card distributions are equally likely). The cards are then turned up one by one from the top until the second ace appears. Prove that the expected (average) number of cards to be turned up is $(n + 1)/2$.

6.5 (U. S. A. Olympiad, 1983) On a given circle, points A, B, C, D, E, and F are chosen at random, independently and uniformly with respect to arc length. Determine the probability that the two triangles ABC and DEF are disjoint, i.e., have no common points.

6.6 (a) A group of n people, X_1, ..., X_n, are throwing a ball back and forth. Everyone throws the ball to everyone else with equal likelihood except that X_1 always throws the ball to X_2. For $1 \leq i \leq n$, let p_i be the probability that X_i has the ball or has just thrown it. Show that

$$(p_1, p_2, \ldots, p_n) = \left(\frac{1}{n}, \frac{2n-2}{n^2}, \frac{n-1}{n^2}, \ldots, \frac{n-1}{n^2} \right).$$

Thus, on average, X_2 has the ball twice as often as every other person except X_1, who has the ball more often than every other person except X_2.

(b) Show that $p_1 = 1/n$ no matter who X_1 throws the ball to (as long as the others throw the ball to everyone with equal likelihood).

6.7 A piece of paper is in the shape of a rectangle $ABCD$ with $AB = CD = 3$ and $AD = BC = 5$. The paper is folded so that A and C coincide. Find the length of the crease.

6.8 (Putnam Competition, 1964) Show that the unit disk in the plane cannot be partitioned into two disjoint congruent subsets.

Solutions

6.1 Rank Amateur plays the two games simultaneously, using the black pieces against Grandmaster One and the white pieces against Grandmaster Two. When Grandmaster One makes his first move, Rank Amateur copies that move with his own white piece on Grandmaster Two's board. When Grandmaster Two responds, Rank Amateur plays the response on Grandmaster One's board. Continuing this way, Rank Amateur plays Grandmaster One's moves on Grandmaster Two's board and *vice versa*. Thus two identical chess games are created, with Rank Amateur playing as White on one board and Black on the other. If Rank Amateur loses one game he wins the other, and if he draws one game then he draws both.

Note: If chess clocks are used, then Rank Amateur must imitate the Grandmasters' moves as soon as they are made, in order to preserve time symmetry.

Note: There is a story that world chess champion Alexander Alekhine was almost defeated by a similar scheme. Two amateurs challenged him to a match in which he would play blindfolded against both of them simultaneously, with the white pieces on one board and the black pieces on the other. The amateurs stipulated that if they earned one point they would win the match. They mimicked Alekhine's moves so that Alekhine was, in effect, playing against himself. Alekhine quickly realized the nature of their trick and played a piece sacrifice that, via some subtle gamesmanship, he convinced them was unsound. The player who should have mimicked the sacrifice chose a different move and went on to lose. The other player was surprised to find out that the sacrifice was sound, as Alekhine won with it, thereby finishing the match 2–0. The act of finding a winning ruse against an apparently impregnable strategy (not to mention winning the games)—all while blindfolded—has to be considered a problem solving feat of the highest order!

6.2 Since
$$f(x) = \frac{f(x) + f(-x)}{2} + \frac{f(x) - f(-x)}{2},$$

such a representation is possible. To show uniqueness, suppose that

$$f(x) = e_1(x) + o_1(x) = e_2(x) + o_2(x),$$

where e_1, e_2 are even and o_1, o_2 are odd. Then

$$f(-x) = e_1(x) - o_1(x) = e_2(x) - o_2(x).$$

It follows that $2e_1(x) = 2e_2(x)$ and hence that $e_1(x) = e_2(x)$ and $o_1(x) = o_2(x)$. This proves the uniqueness of the representation.

6.3 Let E_H and E_T be the events that Alpha obtains more heads and more tails than Beta, respectively. We will show that $\Pr(E_H) = 1/2$. Since Alpha flips more coins than Beta, $\Pr(E_H \cup E_T) = 1$. However, since Alpha has only one coin more than Beta, the events E_H and E_T are mutually exclusive. Therefore $\Pr(E_H) + \Pr(E_T) = 1$. By symmetry, the events are equally likely. Hence $\Pr(E_H) = \Pr(E_T) = 1/2$.

6.4 Suppose that the aces occupy positions a, b, and c in the deck. Then the number of cards turned until the second ace appears is b. However, by symmetry, the deck is equally likely to be in the completely reverse order, in which case the second ace is in the $(n + 1 - b)$th position. Therefore, the expected position of the second ace is

$$\frac{b + (n + 1 - b)}{2} = \frac{n + 1}{2}.$$

6.5 Clockwise from point A, there are 5! permutations of the other five vertices. By symmetry, these permutations are all equally likely. As there are $3! \times 3!$ permutations in which ABC and DEF are disjoint (3! ways to order the vertices of each triangle), the probability that the two triangles are disjoint is $3! \times 3!/5! = 3/10$.

6.6 We first prove (b). It is immediate that

$$p_1 = \frac{1 - p_1}{n - 1},$$

and therefore $p_1 = 1/n$. This proves (b).

Now we prove (a). By symmetry, $p_3 = \cdots = p_n$. Let p denote the common value. Then

$$p = \frac{1}{n - 1}p_2 + \frac{1}{n - 1}(n - 3)p,$$

and it follows that

$$p_2 = 2p.$$

Now,

$$\frac{1}{n} + 2p + (n-2)p = 1$$

and so $p = (n-1)/n^2$. Hence

$$(p_1, p_2, \ldots, p_n) = \left(\frac{1}{n}, \frac{2n-2}{n^2}, \frac{n-1}{n^2}, \ldots, \frac{n-1}{n^2} \right).$$

6.7 Consider the following diagram, where EF is the crease, with E on line BC:

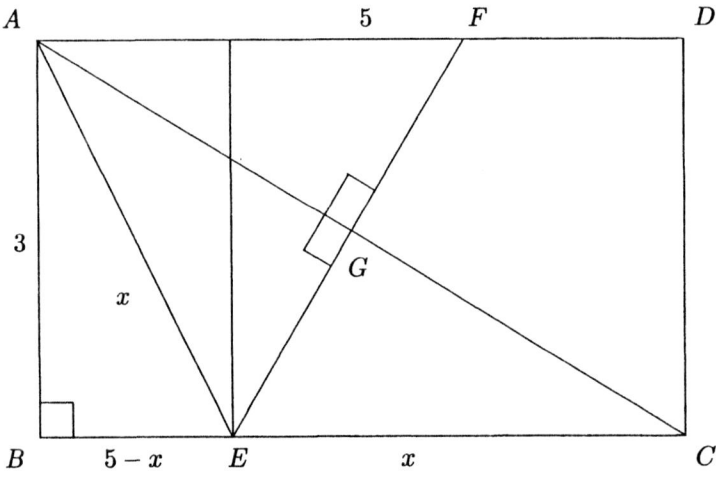

Let x be the length of EC; we calculate x. Since AE and EC are coincident when the paper is folded, they have the same length. Thus

$$3^2 + (5-x)^2 = x^2,$$

and $x = \frac{17}{5}$.

Now we find that the length of EF is

$$2\sqrt{\left(\frac{17}{5}\right)^2 - \left(\frac{1}{2}\sqrt{34}\right)^2} = \frac{3}{5}\sqrt{34}.$$

Note: In general, for a rectangle of width w and length l ($w < l$), the length of the crease is

$$\frac{w}{l}\sqrt{l^2 + w^2}.$$

6.8 If we remove the center of the disk, then the resulting set is the disjoint union of two congruent subsets, namely, two half-disks with half-diameter borders. In the problem as stated, it is the *asymmetric point*, the center, which makes the required decomposition impossible.

Suppose that the unit disk is the disjoint union of two congruent sets A and B. The center O of the disk is in one of these sets. Without loss of generality, suppose that $O \in A$. Since A and B are congruent, there exists a corresponding point, call it O', in B. Construct the ray $O'O$ and suppose that it intersects the circle in point P. Consider the semicircle S perpendicular to $O'O$ and passing through P. All of the points on S belong to A, for any point belonging to B would determine a distance to O' greater than the radius of the circle. However, B does not contain a set of points corresponding to S. Therefore the unit disk is not a disjoint union of two congruent subsets.

Additional Problems

6.9 Suppose that two circles C_1 and C_2 are given on one side of a line l. Show how to construct a point P on l so that there exist tangent lines from P to C_1 and C_2 which form equal angles with l. (In general, there will be four such points.)

Hint: Consider Example 6.1.

6.10 Town A and town B are separated by a river, as in the picture. A road is to be built joining A and B, crossing the river on a bridge. If the bridge must be perpendicular to the river (to minimize costs), and the road is to be as short as possible, find the location of the bridge.

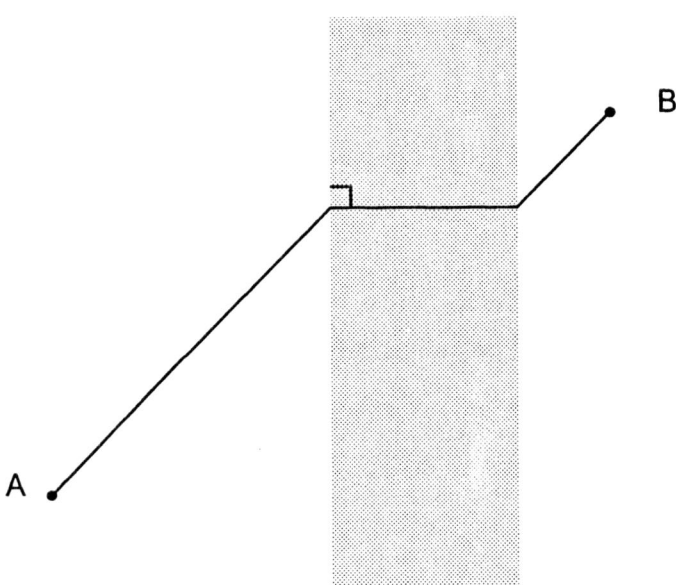

6.11 In the game Knight Packing, two players, Alpha and Beta, alternately place Knights (color unimportant) on a chessboard in such a way that no Knight attacks another Knight. (Once a Knight is placed it cannot be moved.) The last player able to place a Knight on the board is the winner. Who wins Knight Packing? Who wins if Knight Packing is played on a 7 × 7 board?

Note: For the similarly defined game Queen Packing, Alpha wins on an $n \times n$ board for all odd n, but it is not known, in general, who wins if n is even. However, it is not difficult to show by exhaustion that Alpha wins on an 8 × 8 board. In "misère" versions of Knight and Queen Packing, the last player to place the given piece loses. It is not known who wins the misère games for arbitrary values of n.

6.12 Four spheres of radius one are contained in a regular tetrahedron in such a way that each is tangent to three faces of the tetrahedron and to the other three spheres. Find the side length of the tetrahedron.

6.13 Let $ABCD$ be a square with point P in its interior. Line segments are drawn from P to A, B, C, and D. Given that $\angle PAB = \angle PBA = 15°$, show that CDP is an equilateral triangle.

6.14 Evaluate

$$\int_0^{\pi/2} \ln(\sin x)\, dx.$$

6.15 (U. S. A. Olympiad, 1978) $ABCD$ and $A'B'C'D'$ are square maps of the same region of a country but drawn to different uniform scales and are superimposed as shown in the figure. Prove that there is only one point O on the small map which lies directly over a point O' of the large map such that O and O' each represent the same place of the country. Also, give the Euclidean construction (straightedge and compass) for O.

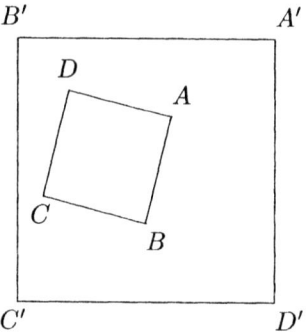

6.16 (*Green Book*, Problem 5) Let $f(x)$ be a continuous function on $[0, a]$, where $a > 0$, such that $f(x) + f(a - x)$ does not vanish on $[0, a]$. Evaluate the integral

$$\int_0^a \frac{f(x)}{f(x) + f(a - x)}\, dx.$$

6.17 Let A, B, C be vertices of a triangle, and let $\overline{A}$, $\overline{B}$, $\overline{C}$ be points on the edges of BC, AC, AB, respectively, positioned so that $\overline{A}C = BC/3$, $\overline{B}A = AC/3$, and $\overline{C}B = AB/3$. Let P be the intersection of $A\overline{A}$ with $B\overline{B}$, Q the intersection of $B\overline{B}$ with $C\overline{C}$, and R the intersection of $C\overline{C}$ with $A\overline{A}$. Show that the area of $\triangle PQR$ is 1/7th of the area of $\triangle ABC$.

6.18 (Putnam Competition, 1998) Given a point (a, b) with $0 < b < a$, determine the minimum perimeter of a triangle with one vertex at (a, b), one on the x-axis, and one on the line $y = x$. You may assume that a triangle of minimum perimeter exists.

6.19 In the game Near–Far, Alpha and Beta each select a point in the unit disc. Alpha wins if the points are within 1/2 of each other and Beta wins if they are not. They play the game repeatedly. What is the best strategy for each of them to adopt, and what is the probability of Alpha winning?

6.20 Given a triangle ABC, show how to construct the point P in the interior that minimizes the quantity $PA + PB + PC$?

Note: P is called the *Fermat point* of the triangle. The Fermat point has an interesting physical interpretation. Suppose that triangle ABC with all angles less than $120°$ is fixed in a plane parallel to and above the ground. A point P is free to move in the interior of the triangle, and to P are attached three strings passing through the vertices A, B, C and hanging down where they hold three equal weights. The Fermat point of the triangle is the point at which P comes to rest. This equilibrium occurs when the potential energy of the system of weights is a minimum, that is, when the sum of the lengths of strings hanging down is a maximum and the sum of the lengths of strings inside the triangle $(PA + PB + PC)$ is a minimum. As a step toward constructing the Fermat point, observe that three force vectors of equal length can only give a net result of zero if their directions differ by $120°$. Thus the Fermat point P has the property that $\angle APB = \angle BPC = \angle CPA = 120°$.

Chapter 7

Parity

"When you're counting alibis and not apples, one plus one equals none."

MARGARET MILLAR
Dr. Paul Prye, *The Weak-Eyed Bat*, 1942

Two integers are said to have the same *parity* if they are both odd or both even. Furthermore, we speak of the parity of an integer as being odd or even. In this chapter we investigate problems that can be solved using the notion of parity.

The main uses of parity in mathematics are via the following simple observations:

1. Two integers of different parity cannot be equal.

2. Adding an odd integer to another integer changes that integer's parity.

3. Adding an even integer to another integer does not change that integer's parity.

In *parity arguments*, the only relevant feature of a number is its parity.

Example 7.1. A magician tosses ten coins onto a table, looks at them, and then faces away from the table. He asks a volunteer to turn over any two of the coins and place a hand over one of them. The magician turns around, looks at the exposed coins, and announces that the covered coin is a "head." The coin is uncovered and it is a head. How is the trick done?

Solution: After tossing the coins, the magician notes whether the number of heads is odd or even. When the volunteer turns two coins over, the number of heads increases by two, decreases by two, or stays the same. In all cases, the *parity* of the number of heads stays the same. When the volunteer covers a coin, the magician looks at the table and observes the number of exposed heads. If the new count agrees in parity with the old count, then the covered coin is a tail. If the parities disagree, then the covered coin is a head.

Note: The trick can be "dressed up" by having the magician ask the volunteer to turn over several pairs of coins rather than just one pair. The result is the same: the parity of the number of heads does not change. ■

Example 7.2. Figure 7.1 shows a closed curve and a point. Is the point inside or outside the curve?

Figure 7.1: A point in a maze.

Solution: We draw a ray from the point and count the number of times it crosses the curve. Each time the ray crosses the curve, it moves from inside the curve to outside or from outside to inside. (This intuitively obvious fact is known as the Jordan curve theorem.) If the number of crossings is even, the point is outside the curve. If the number is odd, the point is inside the curve. In Figure 7.2, for example, the given ray crosses the curve nine times. Hence the point is inside the curve. ■

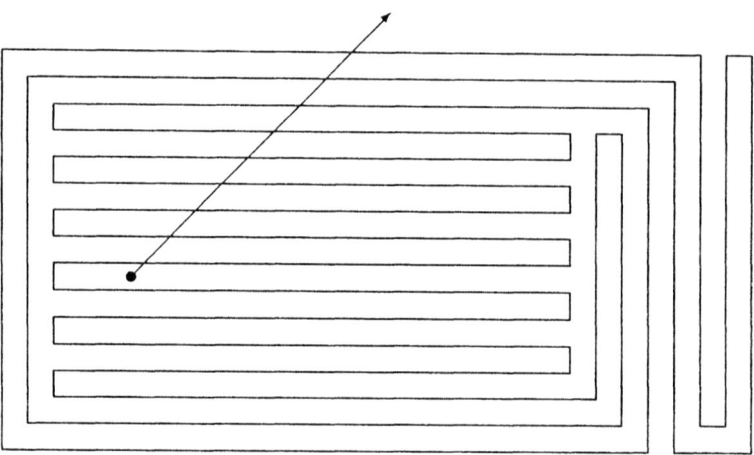

Figure 7.2: A ray crossing the curve nine times.

Example 7.3. Is the matrix

$$A = \begin{bmatrix} 1235 & 2344 & 1234 & 1990 \\ 2124 & 4123 & 1990 & 3026 \\ 1230 & 1234 & 9095 & 1230 \\ 1262 & 2312 & 2324 & 3907 \end{bmatrix}$$

invertible?

Solution: One way to show that a matrix is invertible is to reduce it to row-echelon form and check that the diagonal entries are nonzero. However, for the given matrix A, this procedure would involve cumbersome arithmetical calculations. Instead, we consider the entries of the matrix modulo 2 and things simplify considerably. Let

$$B = \begin{bmatrix} 1 & 0 & 0 & 0 \\ 0 & 1 & 0 & 0 \\ 0 & 0 & 1 & 0 \\ 0 & 0 & 0 & 1 \end{bmatrix}$$

The entries of A and B are the same modulo 2 and therefore so are their determinants. Since the determinant of B is odd, the determinant of A is an odd number and therefore not 0. Hence A is invertible. ∎

Example 7.4. In a certain jail, there are 100 cells, numbered 1 through 100, all locked. During the night, the jailer turns the key on every cell, thereby unlocking them. Then he turns the key on every second cell (2, 4,

6, ...), locking them. Then he turns the key on every third cell (3, 6, 9, ...), and so on, so that on his last pass by the cells, he turns the key on only the 100th cell. Which cells are now open?

Note: This example also appeared as Problem 1.5.

Solution: The open cells are those numbered with perfect squares between 1 and 100, i.e., the numbers 1, 4, 9, 16, 25, 36, 49, 64, 81, and 100. To prove this, we note that a cell is open if and only if its key has been turned an odd number of times, and this happens precisely when the cell number has an odd number of divisors. For each divisor d of an integer n with $d < \sqrt{n}$, there is another divisor d/n with $d/n > \sqrt{n}$. Hence divisors of n occur in pairs, except when $\sqrt{n}$ is an integer. When $\sqrt{n}$ is an integer, i.e., n is a perfect square, there are an odd number of divisors.   ■

Example 7.5. Show that in any graph G with a finite number of vertices (see Glossary), the sum of the degrees of the vertices of G is an even number.

Solution: Each edge contributes two to the sum of the degrees of the vertices of G. Therefore the sum is twice the number of edges of G and hence an even number.

Note: This result is called the "handshake theorem." Let the vertices of the graph represent people at a gathering and the edges signify that the two corresponding people shake hands. The handshake theorem says that the number of handshakes each person participates in, summed over all the people, is an even number.   ■

Problems

7.1 Suppose that one removes two opposite corner squares of a checkerboard. Can the remaining 62 squares be covered with 31 dominoes (of size 2×1)?

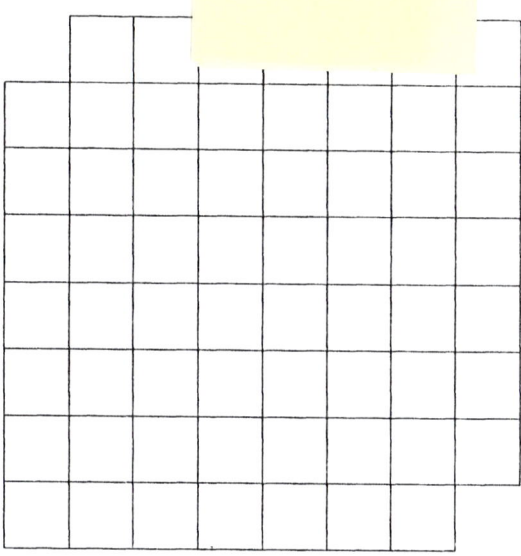

checkerboard with corners removed 2×1 domino

7.2 Suppose that two athletic conferences each have thirteen teams. They want to set up a playing scheduling in which each team plays eleven games against teams in its own conference and three games against teams outside its conference. Is this possible?

7.3 Let a, b, c be odd integers. Show that the quadratic equation

$$ax^2 + bx + c = 0$$

does not have a rational solution.

7.4 Let n be an odd integer greater than 1. Let A be an n by n symmetric matrix such that each row and each column consists of a permutation of the integers $1, \ldots, n$. Show that each integer $1, \ldots, n$ must occur on the main diagonal of A.

Note: An $n \times n$ matrix in which each row and each column contains the numbers $1, \ldots, n$ is called a *Latin square* of order n. Therefore this result shows that in a symmetric Latin square of odd order, the main diagonal is "Latin."

7.5 (Eötvös Competition, 1894; revised) Suppose that $x_1, \ldots, x_n$ is a permutation of the numbers $1, \ldots, n$, with n odd. Show that the product

$$(x_1 - 1)(x_2 - 2) \ldots (x_n - n)$$

is an even number.

7.6 (Wilson's theorem) Show that if p is prime, then $(p-1)! \equiv -1 \bmod p$.

7.7 Show that a group with an even number of elements must contain an element of order two.

7.8 Show that
$$\sum_{k=1}^{n} \frac{1}{k}$$
is never an integer for $n \geq 2$.

7.9 Suppose that 13 stones of integer weight have the property that any 12 of them may be partitioned into two equal weight sets of six elements. Prove that all 13 stones weigh the same.

Solutions

7.1 No. Opposite corner squares of a checkerboard are the same color. Without loss of generality, suppose that they are both black. Removing them, we are left with 32 red and 30 black squares. As each domino covers one red square and one black square, 31 dominoes cannot cover the required squares.

7.2 No such playing schedule exists, for each conference would hold $13 \cdot 11/2$ intra-conference games, and this number is not an integer.

Note: Part of the playing schedule can be modeled by a graph in which the vertices represent the thirteen teams of one conference and the edges represent games within that conference. If such a schedule were possible, the sum of the degrees of the vertices would be 13×11. But this contradicts the "handshake theorem" (Example 7.5), which says that the sum of degrees must be an even number.

7.3 Solution (1): Suppose that $ax^2 + bx + c = 0$ has a rational solution. Then
$$ax^2 + bx + c = (Ax + B)(Cx + D)$$
for some integers A, B, C, and D. Furthermore, $AC = a$, $BC + AD = b$, and $BD = c$. However, since a and c are odd, it follows that A, B, C, and D are all odd, which in turn implies that b is even, a contradiction.

Solution (2): Suppose that $ax^2 + bx + c = 0$ is satisfied by the rational number p/q, with p and q relatively prime. Then
$$ap^2 + bpq + cq^2 = 0,$$

which, modulo 2, becomes

$$p^2 + pq + q^2 = 0.$$

It is easy to check that this equation is not satisified for the ordered pairs $(p, q) = (0, 1)$, $(1, 0)$, and $(1, 1)$. Thus $(p, q) = (0, 0)$, contradicting the fact that p and q are relatively prime.

7.4 Since each row of A contains a permutation of the integers $1, \ldots, n$, each entry i appears an odd number of times in A. Because A is symmetric, the off-diagonal entries occur in pairs. Therefore, there are an odd number of i's on the main diagonal, for each i. Since the main diagonal has n entries, each number i occurs there exactly once. Therefore the main diagonal is a permutation of the integers $1, \ldots, n$.

7.5 This solution uses the pigeonhole principle (Chapter 9) as well as parity. Of the integers $1, \ldots, n$, $(n + 1)/2$ are odd and $(n - 1)/2$ are even. Therefore, in any permutation $x_1, \ldots, x_n$ of these numbers, there exists an odd number i for which x_i is odd. Since $x_i - i$ is even for such an i, it follows that the product

$$(x_1 - 1)(x_2 - 2) \ldots (x_n - n)$$

is an even number.

7.6 Since p is prime, each of the numbers $1, \ldots, p-1$ has a multiplicative inverse modulo p. Each element x is different from its inverse x^{-1}, except for those x which satisfy the congruence $x^2 \equiv 1 \bmod p$, i.e., the elements ± 1. Thus $(p - 1)! \equiv 1 \cdot -1 \equiv -1 \bmod p$.

7.7 Let G be a group of even order. Let S be the set of elements of G which are self-inverses (that is, $x \in S$ if and only if $x = x^{-1}$). Let $T = G - S$. As $|G|$ and $|T|$ are both even, $|S|$ is even. Since the identity element e is a member of S (it satisfies the equation $x = x^{-1}$), S has at least two members. Nonidentity elements of S have order two.

Note: It follows from this result that the product of all the elements in an abelian group of even order is an element $x \neq e$ for which $x^2 = e$. This observation gives another proof of Wilson's theorem (previous problem).

7.8 Let $M_n = \sum_{k=1}^{n} 1/k$. Among the integers $1, \ldots, n$, let m be the largest power of 2. Putting the fractions in the summation over the least common denominator, the numerator of each fraction except

$1/m$ contains a factor of two. Hence the numerator of the sum of the fractions is an odd number (the sum of $n-1$ even numbers and one odd number), while the denominator is even. Hence M_n is not an integer.

7.9 Let the thirteen stones have weights $w_1, \ldots, w_{13}$. When we remove any stone, the other stones can be divided into two groups of equal weight. Therefore the sum of the weights of any twelve stones is even. That is to say, for any distinct i and j, $2s = \sum w_k - w_i$ and $2t = \sum w_k - w_j$. Hence $w_i - w_j$ is even. It follows that all thirteen stones have weights of the same parity.

Let w be the minimum value of the w_i, and let $w_i' = w_i - w$. We must show that $w_i' = 0$ for each i. The w_i' have the same property as the original w_i. Therefore they all have the same parity and in fact they are all even since one of them is 0. If they are not all zero, then let 2^a be the largest power of 2 that divides all the w_i'. Let $w_i'' = w_i'/2^a$. The w_i'' have the same property as the w_i. However, at least one of them is 0 and another is odd, a contradiction. Hence each w_i' is 0, and we are finished.

Additional Problems

7.10 Show that there is no polynomial $p(x)$ with integer coefficients such that $p(1) = 4$ and $p(3) = 5$.

7.11 Is there a choice of a, b, c, d, e, f, g, h, i equal to 1, 2, 3, 4, 5, 6, 7, 8, 9 in some order such that

$$a - b + c - d + e - f + g - h + i = 10 ?$$

7.12 Suppose that 25 students are sitting in a 5×5 array of chairs. Is it possible for them to simultaneously change seats so that each person moves to a neighboring seat in front of, behind, or to the side of his or her original seat?

7.13 Can the five four-cell figures

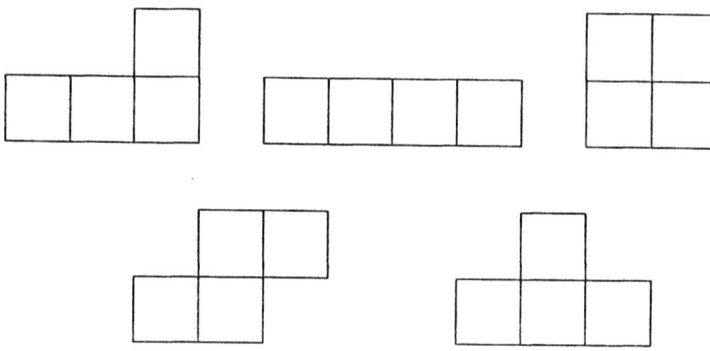

be packed into the 4 × 5 "box"

without overlap? (The figures may be rotated and flipped over if necessary.)

7.14 The members of a cetain club are Truth Tellers (who always make true statements) and Liars (who always make false statements). The members sit at a round table, with Truth Tellers and Liars in alternate seats. One day a woman visits the club and observes the activities. When she returns home, she realizes that she doesn't remember how many club members there are. She phones the club secretary and asks, and the secretary replies, "The club has seventeen members." After this conversation, the visitor realizes that she still doesn't necesssarily know the correct number of club members (the secretary might be a Liar). She phones the president and relates what the secretary told her. The president says, "The secretary is a Liar. There are eighteen members." How many members does the club have?

7.15 Suppose that 32 dominoes tile an 8 × 8 checkerboard. Show that the number of tiles in a vertical orientation must be an even number.

Hint: Instead of using the usual checkerboard coloring, as in Problem 7.1, color the rows alternately with two colors.

7.16 Suppose that n lines are given in the plane. Show that the regions into which the plane is divided can be colored with two colors so that regions sharing a line segment boundary are different colors.

Hint: Choose any region R and color it at random. Define a coloring of the other regions as follows: Given a region R', let l be the number of lines which separate R and R'. If l is even, give R' the same color as R. If l is odd, give R' the other color.

Note: This problem appeared as Additional Problem 4.17.

7.17 Solve the previous problem with 'line' replaced by 'circle'.

Note: This problem appeared as Additional Problem 4.18.

7.18 Let a, b, c, d, e, f, g, h, i be real numbers. Show that in the determinant formula

$$\begin{vmatrix} a & b & c \\ d & e & f \\ g & h & i \end{vmatrix} = aei + bfg + cdh - afh - bdi - ceg,$$

not all six quantities on the right can be positive.

7.19 In the puzzle shown below, fifteen tiles numbered 1, $\ldots$, 15 and a vacant space are arranged in a 4×4 grid. One can move any tile up or down or left or right into the vacant space. Show that one *cannot* transform the configuration on the left into the one on the right by any sequence of moves.

1	2	3	4
5	6	7	8
9	10	11	12
13	14	15	

1	2	3	4
5	6	7	8
9	10	11	12
13	15	14	

Note: One may ask further, which configurations are obtainable from the one on the left?

7.20 An *Euler circuit* of a graph G is a circuit containing all edges of G. Show that if G is connected then G has an Euler circuit if and only if each vertex of G has even degree.

7.21 (*School Science and Mathematics*, Problem 4214, December 1988) Let ABC be a right triangle with integer side lengths, hypotenuse c, semiperimeter s, and area K. Prove that the quantity $(2s - c)s - K$ is a perfect square.

7.22 (Sperner's lemma) Let T be a triangle whose vertices are labeled 1, 2, 3 (a "123 triangle," for short). Some new points are added to the edges and/or the interior of T. Points on edge 12 of T are labeled either 1 or 2; points on edge 13 are labeled 1 or 3; points on edge 23 are labeled 2 or 3; and points in the interior of T are labeled 1, 2, or 3. Finally, T is triangulated into smaller triangles whose vertices are the new points or the vertices of T. Show that one of these smaller triangles is a 123 triangle. (See diagram.)

Hint: A triangle with an odd number of edges labeled 12 must be a 123 triangle.

Note: Brouwer's fixed-point theorem (an important theorem of analysis) follows easily from Sperner's lemma.

(Brouwer's fixed-point theorem) Suppose that f is a continuous map from the unit disc $D \in \mathbf{R}$ into itself. Then f has a fixed point, i.e., a point $p \in D$ such that $f(p) = p$.

Here is a sketch of the proof. We show the result for a triangle (and interior) instead of for a disc. Since a disc and a triangle are homeomorphic, the two figures have the same topological properties. Let the given triangle be ABC. We assign *barycentric coordinates* to all points in the triangle with respect to the vertices A, B, and C. That is, we think of A, B, and C as vectors and coordinatize each point x as $\langle x_1, x_2, x_3 \rangle$, where $x = x_1 A + x_2 B + x_3 C$, $x_1 + x_2 + x_3 = 1$, and $x_i \geq 0$ for $i = 1$, 2, 3. Notice that A, B, and C have coordinates $\langle 1, 0, 0 \rangle$, $\langle 0, 1, 0 \rangle$, $\langle 0, 0, 1 \rangle$, respectively. Now we partition ABC into many small triangles, and we label the vertices of these small triangles as follows. Suppose that vertex $y = \langle y_1, y_2, y_3 \rangle$ is sent to $y' = \langle y_1', y_2', y_3' \rangle$. Then y is labeled i, $i = 1$, 2, 3, if $y_i \geq y_i'$. Notice that A is labeled 1, B is labeled 2, and C is labeled 3, and, furthermore, that each vertex must have some label but could have two or even three labels. By Sperner's lemma, there must be a small 123 triangle, say, T. Repeating this process, we obtain an infinite sequence of arbitrarily small, nested 123 triangles:

$$ABC \supseteq T \supseteq T' \supseteq T'' \supseteq \cdots.$$

Since triangles (with interiors) are compact sets, there must be a point p contained in all the triangles of this sequence. We claim that p is a fixed point of f. As p is arbitrarily close to points labeled 1, points labeled 2, and points labeled 3, p is labeled 1, 2, and 3. And any point labeled 1, 2, 3 must be a fixed point, since

$$1 = y_1 + y_2 + y_3 \geq y_1' + y_2' + y_3' = 1$$

and $y_1 \geq y_1'$, $y_2 \geq y_2'$, $y_3 \geq y_3'$ imply that $y_1 = y_1'$, $y_2 = y_2'$, and $y_3 = y_3'$.

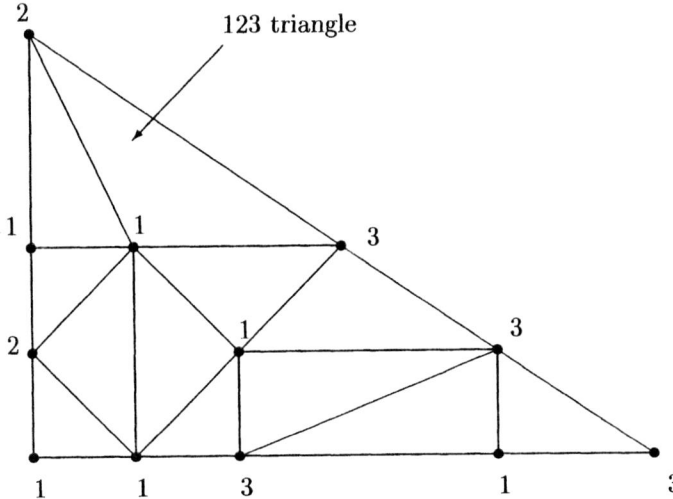

Chapter 8

Various Moduli

"Sometimes a detective has to make two and two into six, at the very least."

STUART PALMER
Miss Hildegarde Withers,
The Puzzle of the Silver Persian, 1934

In the previous chapter we discussed problems whose solutions involved reducing all numbers to one of two types, odd or even. Such parity arguments are also called "modulo 2 arguments." In this chapter, we generalize the procedure to "modulo n arguments," where n is a suitable value.

We start with a simple problem.

Example 8.1. Show that $n^2 + 1$ is divisible by 7 for no positive integer n.

Solution: We test all possible values of n modulo 7: $0^2 + 1 \equiv 1$, $1^2 + 1 \equiv 2$, $2^2 + 1 \equiv 5$, $3^2 + 1 \equiv 3$, $4^2 + 1 \equiv 3$, $5^2 + 1 \equiv 5$, and $6^2 + 1 \equiv 2$. As we never obtain the residue 0, $n^2 + 1$ is divisible by 7 for no positive integer n. (In fact, we really only needed to test the values $n = 0$, 1, 2, and 3, since $(-n)^2 = n^2$.) ∎

In the above problem, we found that although $n^2 + 1$ never takes the value 0 modulo 7, it does take four other values. Of course, n^2 takes four values also. Things are much simpler modulo 4, where n^2 has only two possible values. Always,

$$n^2 \equiv 0 \text{ or } 1 \bmod 4.$$

To show this, we just compute the squares of the different residue classes modulo 4: $0^2 \equiv 0$, $1^2 \equiv 1$, $2^2 \equiv 0$, and $3^2 \equiv 1$.

This is a handy fact, as will now see.

Example 8.2. Show that the equation

$$x^2 + y^2 = 1003$$

has no integer solutions.

Solution: By the above fact,

$$x^2 + y^2 \equiv 0, 1, \text{ or } 2 \bmod 4.$$

Since $1003 \equiv 3 \bmod 4$, there are no integer solutions to the given equation.

Note: A relation such as the one above, in which we allow only integer values for the variables, is called a *Diophantine equation.*   ■

The situation of squares modulo 8 is also particularly simple. Always,

$$n^2 \equiv 0, 1, \text{ or } 4 \bmod 8.$$

(This is shown by computing the squares of the different residue classes modulo 8.)

Example 8.3. Show that no integer of the form $8n + 7$ is the sum of three squares.

Solution: By the above fact,

$$x^2 + y^2 + z^2 \equiv 0, 1, 2, 3, 4, 5, \text{ or } 6 \bmod 8.$$

Therefore, $x^2 + y^2 + z^2$ is never equal to an integer of the form $8n + 7$.

Note: A famous theorem of Lagrange asserts that every positive integer is the sum of *four* squares.   ■

A similar situation exists for squares modulo 16. We have

$$n^2 \equiv 0, 1, 4, \text{ or } 9 \bmod 16.$$

(Again, just compute the squares of the various residue classes modulo 16.)
    The situation for the modulus 32 and other higher powers of 2 is not as simple.

    The following result is useful in dealing with Diophantine equations in which the exponents are higher powers of two.

Theorem 8.1. For all $n \geq 2$,

$$x^{2^n} \equiv 0 \text{ or } 1 \bmod 2^{n+2}.$$

Specifically, if x is even, then

$$x^{2^n} \equiv 0 \bmod 2^{n+2},$$

and if x is odd, then

$$x^{2^n} \equiv 1 \bmod 2^{n+2}.$$

Proof. If x is even, say, $x = 2k$, then $2^n \geq n + 2$ (since $n \geq 2$) and

$$x^{2^n} = (2k)^{2^n} = 2^{2^n} k^{2^n} \equiv 0 \bmod 2^{n+2}.$$

For x odd, we proceed by induction on n. Let $x = 2k + 1$. The case $n = 2$ says that

$$x^4 \equiv 1 \bmod 16,$$

which is verified as follows:

$$x^4 = 16k^4 + 32k^3 + 24k^2 + 8k + 1 \equiv 8k(k + 1) + 1 \equiv 1 \bmod 16.$$

Now assume that the result is true for n. Then

$$x^{2^n} = 2^{n+2}m + 1.$$

Thus

$$x^{2^{n+1}} = (2^{n+2}m + 1)^2 = 2^{2n+4}m^2 + 2^{n+3}m + 1 \equiv 1 \bmod 2^{n+3}.$$

∎

Example 8.4. (U. S. A. Olympiad, 1979) Determine all non-negative integral solutions $(n_1, \ldots, n_{14})$ if any, apart from permutations, of the Diophantine equation $n_1^4 + \cdots + n_{14}^4 = 1599$.

Solution: The 4th powers lead us to consider a modulo 16 argument, and the presence of the number 1599 seems to confirm our thinking. By Theorem 8.1, $n_1^4 + \cdots + n_{14}^4 \equiv 0, 1, 2, \ldots, 13$, or 14 mod 16. But $1599 \equiv 15 \bmod 16$. Therefore the equation has no solutions. ∎

Problems

8.1 Show that a positive integer n is divisible by 3 if and only if the sum of its digits is divisible by 3.

8.2 Suppose that a sequence $a_0, a_1, a_2, \ldots$ of positive integers is defined recursively by $a_n = 2a_{n-1} + 1$. For which values of a_0 are infinitely many integers of the sequence divisible by 5?

8.3 (Eötvös Competition, 1894) Prove that the expressions $2x + 3y$ and $9x + 5y$ are divisible by 17 for the same set of integral values of x and y.

8.4 Prove that the equation

$$(3n^2 + 1)^{100} + (-3n^2 + 1)^{99} = 4$$

holds for no integer n.

8.5 (Putnam Competition, 1954) Prove that there are no integers x and y for which
$$x^2 + 3xy - 2y^2 = 122.$$

Solutions

8.1 Let $n = \sum_{k=0}^{m} a_k 10^k$. Then

$$n = \sum_{k=0}^{m} a_k (9 + 1)^k \equiv \sum_{k=0}^{m} a_k \bmod 3.$$

8.2 Computed modulo 5, the function $a_n = 2a_{n-1} + 1$ maps 0 to 1, 1 to 3, 3 to 2, and 2 to 0. Therefore, if $a_0 \equiv 0$, 1, 2, or 3, then $a_n \equiv 0$ infinitely often. However, if $a_0 \equiv 4$, then $a_1 \equiv 4$, so $a_n \equiv 4$ for all n and $a_n \equiv 0$ for no n.

8.3 Since
$$-5(2x + 3y) + 3(9x + 5y) = 17x,$$

it follows that $17 \mid 2x + 3y$ if and only if $17 \mid 9x + 5y$.

8.4 The left side of the equation is congruent to 2 modulo 3, but the right side is congruent to 1. Hence there are no integer solutions.

8.5 There are no solutions to the equation

$$x^2 + 3xy - 2y^2 = 122$$

in integers. Multiplying the equation by 4 and completing a square we obtain
$$(2x + 3y)^2 - 17y^2 = 488.$$

The presence of   ts considering the equation
modulo 17. Thu

$$(2x + 3y)^2 \equiv 488 \equiv 12 \bmod 17.$$

This congruence requires that 12 be a square modulo 17, but it is easy to check by exhaustion that it is not.

Additional Problems

8.6 (a) What day of the week will it be 365 days from Monday?

(b) Show that there is at least one Friday the thirteenth each year. What is the maximum possible number of Friday the thirteenths in a year?

8.7 You write a 10-digit number and beneath it the same ten digits but in a different order. You subtract the smaller number from the larger and circle one of the nonzero digits in the difference. You read out all the uncircled digits in any order to a friend. Your friend (who is not looking at your numbers) then immediately tells you what the circled digit is. How does your friend do this?

8.8 Show that the product of the side lengths of a Pythagorean triangle is divisible by 60.

Note: See the solution to Problem 1.8.

8.9 (U. S. A. Olympiad, 1976) Determine (with proof) all integral solutions of $a^2 + b^2 + c^2 = a^2b^2$.

8.10 (U. S. A. Olympiad, 1973; modified) Let $\{X_n\}$ and $\{Y_n\}$ denote two sequences of integers defined for as follows:

$$X_0 = 1, \quad X_1 = 1, \quad X_{n+1} = X_n + 2X_{n-1}, \quad n \geq 1$$

$$Y_0 = 1, \quad Y_1 = 7, \quad Y_{n+1} = 2Y_n + 3Y_{n-1}, \quad n \geq 1.$$

Thus, the first few terms of the sequence are:

$$X \quad : \quad 1, 1, 3, 5, 11, 21, \ldots,$$

$$Y \quad : \quad 1, 7, 17, 55, 161, 487, \ldots.$$

Prove that no term other than 1 occurs in both sequences.

Hint: Try a modulo 8 argument.

8.11 Find all integral solutions to the equation

$$a^2 + 5b^2 = 2c^2 + 2cd + 3d^2.$$

Hint: Try a modulo 5 argument.

8.12 Show that it is impossible to tile a square of side length 25 using squares of side length 2 and squares of side length 3.

8.13 (*Green Book*, Problem 8) Prove that the equation

$$x^4 + y^4 + z^4 - 2y^2z^2 - 2z^2x^2 - 2x^2y^2 = 24$$

has no solutions in integers x, y, z.

8.14 (*American Math. Monthly*, Problem E3269, 1988) For what positive integers n does there exist a permutation $(x_1, x_2, \ldots, x_n)$ of $(1, 2, \ldots, n)$ such that the differences $|x_k - k|$, $1 \leq k \leq n$, are all distinct?

8.15 (*American Math. Monthly*, Problem E3313, March 1989) Prove that the number of terms in the expansion of $(x_1 + x_2 + \cdots + x_m)^n$ whose coefficients are not divisible by a given prime number p is

$$\prod_{i=0}^{k} \binom{a_i + m - 1}{m - 1},$$

where $n = \sum_{j=0}^{k} a_j p^j$, $0 \leq a_i < p$, is the p-adic expansion of the positive integer n.

8.16 (Putnam Competition, 1997) For each positive integer n write the sum $\sum_{m=1}^{n} \frac{1}{m}$ in the form p_n/q_n where p_n and q_n are relatively prime positive integers. Determine all n such that 5 does not divide q_n.

Note: Compare with Problem 7.8.

8.17 (*Math. Magazine*, Problem 1297, June 1988) For k a positive integer, define A_n for $n = 1, 2, \ldots$ by

$$A_{n+1} = \frac{nA_n + 2(n + 1)^{2k}}{n + 2},$$

where $A_1 = 1$. Prove that A_n is an integer for all $n \geq 1$, and A_n is odd if and only if n is congruent to 1 or 2 modulo 4.

Chapter 9

Pigeonhole Principle

"I don't like crooks, and even if I did, I wouldn't like crooks that are stool-pigeons, and if I liked crooks that are stool-pigeons, I still wouldn't like you."

DASHIELL HAMMETT
Miriam Nunheim, *The Thin Man*, 1934

The pigeonhole principle says that if many objects are put into a few categories, then fairly many objects must be in one category. This is specified as follows.

Theorem 9.1 (Pigeonhole principle). If $kn + 1$ objects are placed in n pigeonholes, then some pigeonhole contains at least $k + 1$ objects.

Proof. If each pigeonhole contains at most k objects, then the total number of objects is at most nk, a contradiction. Therefore some pigeonhole contains at least $k + 1$ objects. ∎

Example 9.1. Prove that in a theater with 1000 people, some three people share the same birthday.

Solution: The result follows from Theorem 9.1 with $k = 2$ and $n = 366$. Let the pigeonholes be the 366 days of the year (including February 29). By the theorem, if the theater contains $2 \cdot 366 + 1 = 733$ people, then some three share the same birthday. Thus we have the required result with 267 people to spare. ∎

Example 9.2. Suppose that 17 playing cards are chosen at random from a standard deck of 52 cards. Show that these cards must contain a "flush" (five cards all of the same suit).

Solution: The result follows from the pigeonhole principle as stated above, with $k = 4$ and $n = 4$. Let the pigeonholes be the four suits: clubs, diamonds, hearts, and spades. Given 17 cards, some pigeonhole contains at least five cards; i.e., there is a flush in that suit.

Note: Seventeen cards do not necessarily include a "straight" (five cards of consecutive rank). For example, the cards might be the 2's, 3's, 4's, and 5's of all four suits and the 7 of clubs. Let f_n and s_n be the probabilities that n cards chosen at random from a standard deck contain, respectively, a flush and a straight. Poker players know that $f_5 < s_5$ (that is why a flush beats a straight). However, we have shown that $f_{17} > s_{17}$. For which values of n, $5 \leq n \leq 52$, does the inequality $f_n < s_n$ hold?   ∎

The special case of the pigeonhole principle in which $k = 1$ is worth mentioning separately.

Theorem 9.2. If $n + 1$ objects are placed in n pigeonholes, then some pigeonhole contains at least two objects.

Example 9.3. A *lattice point* in the plane is an ordered pair $p = (x, y)$ with integer coordinates x and y. Given five lattice points in the plane, p_1, p_2, p_3, p_4, p_5, show that the midpoint of the line segment $p_i p_j$ determined by some two distinct lattice points p_i and p_j is also a lattice point. See Figure 9.1.

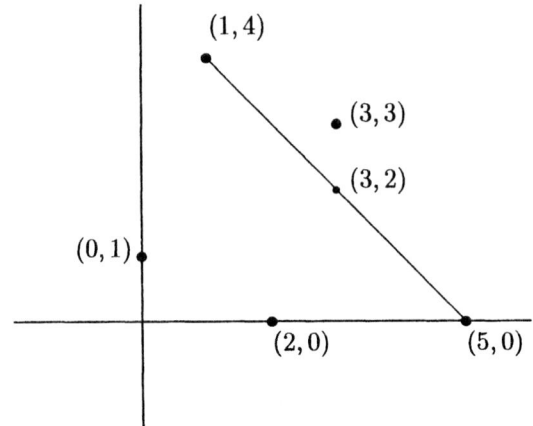

Figure 9.1: Five lattice points and a lattice midpoint.

Solution: There are four possibilities for the parities of an ordered pair of integers: (even, even), (even, odd), (odd, even), and (odd, odd). Therefore, in any set of five lattice points, there exist two points which agree in parity in both coordinates. Such points determine a lattice midpoint, as the midpoint of the line segment joining them is given by the ordered pair of coordinate averages, and if two numbers agree in parity, then their sum is even and their average is an integer. ∎

Example 9.4. Show that in any graph G with a finite number of vertices, some two vertices have the same degree.

Solution: Suppose that G has p vertices. Then each vertex has degree equal to one of the numbers $0, \ldots, p-1$. However, it is impossible for G to have both a vertex of degree 0 and a vertex of degree $p-1$. Therefore the degrees of the p vertices consist of at most $p-1$ different numbers. By the pigeonhole principle, some two vertices have the same degree. ∎

Example 9.5. Suppose that G is a graph with $2n$ vertices and n^2+1 edges, $n \geq 2$. Show that G contains a triangle (three vertices all of which are connected).

Note: The complete bipartite graph $K_{n,n}$, consisting of two disjoint sets of n vertices each and all the edges between the two sets, has n^2 edges and no triangle. Hence, the result stated above is "best possible."

Solution: The proof is by induction on n. The case $n = 2$ is immediate: a graph with four vertices and five edges contains a triangle. For $n \geq 2$, assume that every graph with $2n$ vertices and n^2+1 edges contains a triangle. Let G be a graph with $2(n+1)$ vertices and $(n+1)^2+1$ edges. Let x and y be adjacent vertices in G, and let V be the set of other vertices of G. If G contains more than n^2 edges between vertices of V, then we are finished by the induction hypothesis. Assume that there are at most n^2 edges between vertices of V. Therefore there are at least $2n+1$ edges from x and y to elements of V. By the pigeonhole principle, there is an edge from x and an edge from y to some vertex $z \in V$. Therefore G contains the triangle xyz. Hence every graph with $2n$ vertices and n^2+1 edges contains a triangle. ∎

Problems

9.1 (Putnam Competition, 1978; modified) Let A be any set of 19 distinct integers chosen from the arithmetic progression

$$1, \ 4, \ 7, \ \ldots, \ 100.$$

Prove that there must be two distinct integers in A whose sum is 104.

9.2 Using the pigeonhole principle, show that some positive integral power of 17 ends in 0001 (base ten).

9.3 Suppose that $|S| = 101$, $S \subseteq \{1, \ldots, 200\}$. Show that there are three elements x, y, and z of S for which $x + y = z$.

9.4 Let n points be given in the plane. Prove that three of them form an angle which is at most π/n.

9.5 Given any real $t_1, \ldots, t_7$, prove that there exist i, j with $1 \leq i, j \leq 7$ and $i \neq j$ such that $1 + t_i t_j \neq 0$ and

$$0 \leq \frac{t_i - t_j}{1 + t_i t_j} < \frac{1}{\sqrt{3}}.$$

9.6 Suppose that $a_1, a_2, a_3, \ldots, a_{n+1}$ are positive integers not exceeding $2n$. Prove that there are subscripts i and j such that a_i is a multiple of a_j (and $i \neq j$).

9.7 (*Green Book*, Problem 54; modified) Let $a_1, a_2, \ldots, a_{44}$ be natural numbers such that $0 < a_1 < a_2 < \cdots < a_{44} \leq 125$. Prove that at least one of the differences $d_i = a_{i+1} - a_i$, $i = 1, \ldots, 43$, occurs at least ten times.

9.8 Let S be a set of ten positive numbers less than 100. Show that S contains two disjoint subsets A and B such that the sum of the elements of A is the same as the sum of the elements of B.

9.9 If $a_1, \ldots, a_n$ are any integers, show that there exist indices i and j, $1 \leq i \leq j \leq n$, such that $a_i + a_{i+1} + \cdots + a_j$ is a multiple of n.

9.10 Let q be an odd integer greater than 1. Show that there is a positive integer n such that $2^n - 1$ is a multiple of q.

9.11 A square-free number is one not divisible by the square of any prime.

(a) Given that

$$\sum_{p \text{ prime}} p^{-2} < 0.46,$$

prove that for any positive integer n, more than 54% of the integers $1, \ldots, n$ are square-free.

(b) Prove that every integer greater than 1 is the sum of two square-free positive integers.

9.12 For $n \geq 1$, let $t_0, \ldots, t_{n-1}, t_n = t_0$ be real numbers with $t_i < t_{i-1} + 1$ ($1 \leq i \leq n$). Prove that there exists i with $1 \leq i \leq n$ such that the interval $[t_i, t_{i-1} + 1)$ contains an integer congruent to i modulo n.

9.13 Show that every polyhedron has two faces with the same number of edges.

9.14 Let A be an $m \times n$ matrix with distinct real number entries such that each row's entries are in increasing order from left to right. Form a new matrix A' by arranging the elements of each column of A in increasing order from top to bottom. Show that the elements of each row of A' are in increasing order from left to right.

Note: This situation is known as the "marching band problem." Suppose that the players in a marching band are arranged in a rectangle with the members of each row ordered from shortest to tallest. The members of each column now arrange themselves from shortest to tallest. Our result says that after this rearrangement the members of each row are still arranged from shortest to tallest (although people may have changed rows).

9.15 Suppose that six circles (of possibly different radii) are arranged in the plane so that no one of them contains the center of another. Prove that they cannot have a point in common.

9.16 Let C be a circle of radius 16, and let A be an annulus of outer radius 3 and inner radius 2. Let S be a set of 650 points in the interior of C. Show that some copy of A covers at least 10 points of S.

Solutions

9.1 Let the pigeonholes be the sets $\{1\}$, $\{4, 100\}$, ..., $\{49, 55\}$, $\{52\}$. By the pigeonhole principle, some two of the chosen integers must be in the same pigeonhole, i.e., one of the two-element sets of the form $\{a, 104 - a\}$. The elements of such a set sum to 104.

9.2 Consider 17^k, for $k = 1$, ..., 10001. By the pigeonhole principle, some two of these powers, say 17^x and 17^y ($x > y$), agree modulo 10000. It follows that

$$17^{x-y} \equiv 1 \bmod 10000.$$

We observe that '17' is a red herring. The same result is true for any number a with $\gcd(a, 10) = 1$.

Note: The above solution is "nonconstructive" in the sense that, although it shows the existence of an exponent that satisfies the congruence, it doesn't furnish one. However, by Euler's theorem (Glossary), $17^{\phi(10000)} \equiv 1 \bmod 10000$, so we have a solution, namely, $\phi(10000) = 4000$.

9.3 Let z be the largest element in S. We will show that S contains two elements x, y such that $x + y = z$. Consider the sets

$$\{1, z - 1\}, \{2, z - 2\}, \ldots.$$

As $z \leq 200$, there are at most 99 of these sets. Therefore S must contain two elements x, y which lie in the same set. These x, y satisfy the equation $x + y = z$.

9.4 Let the n points be labeled as $p_1, p_2, \ldots, p_n$. Note that if any three of the points are collinear then an angle of measure zero exists, so we may assume that no three of them are collinear. Let P be the convex polygon obtained as the convex hull of the n points. Then each of the points is either a vertex or an interior point of P and at least three of them are vertices. Assume then that p_1, p_2, p_n are adjacent vertices of P with p_1 "between" p_2 and p_n. Then (by relabeling points as needed) we have

$$
\begin{aligned}
\pi &= m(\angle p_1 p_2 p_n) + m(\angle p_1 p_n p_2) + m(\angle p_2 p_1 p_n) \\
&= m(\angle p_1 p_2 p_n) + m(\angle p_1 p_n p_2) + \sum_{i=2}^{n-1} m(\angle p_i p_1 p_{i+1}).
\end{aligned}
$$

So by the pigeonhole principle, at least one of these n angles has measure at most π/n.

9.5 Let $\theta_i = \tan^{-1} t_i$, where we choose θ_i with $-\pi/2 < \theta_i < \pi/2$. By the pigeonhole principle, there exist θ_i and θ_j with $0 \leq \theta_i - \theta_j < \pi/6$. Since $\cos(\theta_i - \theta_j) \neq 0$, it follows that $\cos\theta_i \cos\theta_j + \sin\theta_i \sin\theta_j \neq 0$. Hence $1 + \tan\theta_i \tan\theta_j \neq 0$, i.e., $1 + t_i t_j \neq 0$. As

$$\frac{t_i - t_j}{1 + t_i t_j} = \frac{\tan\theta_i - \tan\theta_j}{1 + \tan\theta_i \tan\theta_j} = \tan(\theta_i - \theta_j),$$

the required inequality is satisfied by t_i and t_j.

9.6 The given integers may be expressed in the form $a_j = 2^{n_j} \cdot b_j$, where each n_j is a nonnegative integer and each b_j is an odd positive integer. Since there are only n odd integers from 1 to $2n$, there must exist $i \neq j$ for which $b_i = b_j$, and then either a_j/a_i (if $n_j > n_i$) or a_i/a_j (if $n_i > n_j$) is an integer.

9.7 Suppose that none of the differences $d_i = a_{i+1} - a_i$ occurs ten times. Note that

$$\sum_{i=1}^{43} d_i = a_{44} - a_1 \leq 125 - 1 = 124.$$

However, the least possible values for these d_i are nine 1's, nine 2's, nine 3's, nine 4's, and seven 5's, totaling 125 (a contradiction). Hence some difference d_i occurs at least ten times.

9.8 Note that the maximum possible sum of elements in S is

$$100 + 99 + 98 + 97 + 96 + 95 + 94 + 93 + 92 + 91 = 955.$$

As there are $2^{10} - 1 = 1023$ different nonempty subsets of S, by the pigeonhole principle, some two subsets of S have the same sum. Call these subsets X and Y. Let $A = X - X \cap Y$ and $B = Y - X \cap Y$. Clearly A and B are disjoint subsets of S. Since we have removed the same elements from X and Y, the sum of the elements of A is the same as the sum of the elements of B.

9.9 The result is clear if one of the n integers

$$a_1, \ a_1 + a_2, \ a_1 + a_2 + a_3, \ \ldots, \ a_1 + \cdots + a_n$$

is a multiple of n. Otherwise, since there are $n-1$ nonzero congruence classes modulo n, some two of these "partial sum" integers must be congruent, say

$$a_1 + a_2 + \cdots + a_{i-1} + a_i + \cdots + a_j \equiv a_1 + a_2 + \cdots + a_{i-1} \bmod n,$$

and therefore $a_i + a_{i+1} + \cdots + a_j \equiv 0 \bmod n$.

9.10 If $2^n \equiv 1 \bmod q$ for some n, $1 \le n \le q$, then $2^n - 1 \equiv 0 \bmod q$ as needed. Otherwise, since there are $q - 1$ "non-one" equivalence classes modulo q, there must exist i and j with $1 \le i < j \le q$ such that $2^i \equiv 2^j \bmod q$. Since 2 and q are relatively prime, we can cancel 2^i to obtain $2^{j-i} \equiv 1 \bmod q$. Hence $2^{j-i} - 1$ is a multiple of q.

9.11 (a) Let q be the largest prime for which $q^2 \le n$. For each prime p with $2 \le p \le q$, the $\lfloor \frac{n}{p^2} \rfloor$ integers p^2, $2p^2$, $\ldots$, $\lfloor \frac{n}{p^2} \rfloor p^2$ are non-square-free, and as p ranges, all non-square-free integers from 1 to n are obtained (with many repetitions). Let $f(n)$ be the number of non-square-free integers not larger than n. Then

$$f(n) < \sum_{p \le n} \left\lfloor \frac{n}{p^2} \right\rfloor \le \sum_{p \le n} \frac{n}{p^2} < 0.46n,$$

and so the number of square-free integers is larger than $0.54n$.

(b) Consider the decomposition $n = j + (n - j)$, $1 \le j \le n$. Since more than 54% of the j values are square-free and also more than 54% of the $n - j$ values are square-free, it follows by the pigeonhole principle that for some j, both j and $n - j$ are square-free.

9.12 For $1 \leq i \leq n$, shift the ith interval, $[t_i, t_{i-1} + 1)$, i units to the left. This produces the pairwise disjoint intervals

$$[t_1 - 1, t_0), \; [t_2 - 2, t_1 - 1), \; \ldots, \; [t_{n-1} - n + 1, t_{n-2} - n + 2),$$

$$[t_0 - n, t_{n-1} - n + 1),$$

whose union is the interval $[t_0 - n, t_0)$. Since $[t_0 - n, t_0)$ has length n, it must contain an integer N with $N \equiv 0 \bmod n$, and hence there must exist an i for which $N \in [t_i - i, t_{i-1} - i + 1)$, and hence $N + i \in [t_i, t_{i-1} + 1)$.

9.13 Let F be a face with the maximum number of edges, n. Then, since F is bounded by n other faces, the total number of faces is at least $n + 1$. As each face must be bounded by some number of edges from the set 3, ..., n, it follows from the pigeonhole principle that some two faces are bounded by the same number of edges.

Note: This result is also implied by the fact that every finite graph has two vertices of the same degree (Example 9.4). For let each face of the polyhedron be represented by a vertex and let two vertices be adjacent when the corresponding faces of the polyhedron are adjacent.

One can show, in fact, that there are two pairs of faces bounded by the same number of edges.

9.14 Suppose that in some row of A' there are two elements, x and y, with $x > y$ and x appearing to the left of y. Consider the partial column X of A' consisting of x and all entries above x and the partial column Y consisting of y and all entries below y. These two partial columns have altogether $m + 1$ elements. Therefore, by the pigeonhole principle, some two of these elements, say $a \in X$ and $b \in Y$, are in the same row of A. But then $a \leq b$, which contradicts the fact that $a \geq x > y \geq b$ in A'.

9.15 Suppose that the six circles, with centers c_1, ..., c_6, have a point p in common. Then, by the pigeonhole principle, $m(\angle c_i p c_j) \leq \pi/3$, for some c_i and c_j. Let the radii of the circles with centers c_i and c_j be r_i and r_j, respectively. Without loss of generality, suppose that $r_j \geq r_i$. Let the distance between c_i and p be a, the distance between c_j and p be b, and the distance between c_i and c_j be c. Then, by the law of cosines,

$$c^2 \;=\; a^2 + b^2 - 2ab \cos m(\angle c_i p c_j)$$

$$\leq \quad a^2 + b^2 - 2ab\frac{1}{2}$$
$$\leq \quad a^2 + b^2 - \min\{a^2, b^2\}$$
$$= \quad \max\{a^2, b^2\}$$
$$\leq \quad r_j^2.$$

Therefore, $c \leq r_j$. Hence the circle with center c_j contains c_i, which contradicts the assumption that no circle contains the center of another. Therefore there is no point common to all six circles.

9.16 Place a copy of the annulus A with center at each of the 650 points. All the annuli lie inside the circle concentric with C with radius 19. Some point of this larger circle must be covered by 10 of these annuli, for if not then the area of the annuli is at most nine times the area of the larger circle, but

$$650 \cdot 5\pi > 9 \cdot 361\pi.$$

Let P be a point covered by 10 of the annuli. The centers of these annuli lie inside an annulus centered at P.

Additional Problems

9.17 (Putnam Competition, 1971) Let there be given nine lattice points (points with integer coordinates) in three-dimensional Euclidean space. Show that there is a lattice point on the interior of one of the line segments joining two of these points.

9.18 Given any five points inside an equilateral triangle with side length 2, show that there are two points at most 1 unit apart.

9.19 Suppose that there are 101 points in the plane with the property that, of any three, some two are less than 1 unit apart. Show that there is a circle of radius 1 containing at least 51 of the points.

9.20 Given any 51 points in a square of side length 7, show that some three are contained in a circle of radius 1.

9.21 Suppose that a_1, a_2, a_3, $\ldots$, a_{n+1} are distinct positive integers not exceeding $2n$. Prove that there are subscripts i and j such that a_i and a_j are relatively prime.

9.22 Suppose that a and b are distinct positive integers and S is a subset of $\{1, 2, \ldots, a+b\}$ with $|S| > (a+b)/2$. Show that some two integers in S differ by a or b.

9.23 (Putnam Competition, 1990) Prove that any convex pentagon whose vertices (no three of which are collinear) have integer coordinates must have area $\geq 5/2$.

Hint: Use the result of Example 9.3 and Pick's theorem (Glossary).

9.24 Show that if the fraction a/b is expressed as a decimal number, the decimal is either terminating or repeating with a period of length no longer than $b - 1$.

9.25 Let G be a graph with n vertices, with each vertex having degree greater than $n/2$. Show that G is connected (i.e., it is possible to get from any vertex to any other vertex by traveling along edges).

Note: See Additional Problem 14.11.

9.26 Suppose that $a > 1$ and $2^b > a + 1$. Given any set S of b integers, show that there are two subsets of S, the sum of whose elements is the same modulo a. Show that this result is not true if $2^b = a + 1$.

9.27 Show that nine checkers (color unimportant) cannot be placed on an 8×8 checkerboard so that all distances between pairs of checkers are different.

9.28 (*Math. Magazine*, Problem 1046, May 1978) For an arbitrary positive integer k, consider the decimal integer h consisting of m copies of k followed by n zeros. Show that for each positive integer x, there exists an m, $m \neq 0$, and an n such that x divides h.

9.29 (Erdös–Szekeres theorem) Show that every sequence of $n^2 + 1$ distinct real numbers contains an increasing subsequence of $n + 1$ terms or a decreasing subsequence of $n + 1$ terms (or both).

9.30 Show that for all real numbers α and $n \in \mathbf{Z}^+$, there exist positive integers p and q with $1 \leq q \leq n$ and

$$\left| \alpha - \frac{p}{q} \right| < \frac{1}{qn} \leq \frac{1}{q^2}.$$

9.31 Let A and B be two nonempty sets of residue classes (modulo n) containing α and β classes, respectively. Show that if $\alpha + \beta > n$, then for any integer k there exist $a \in A$ and $b \in B$ such that $a + b \equiv k \bmod n$.

9.32 (*Math. Magazine*, Problem 1152, September 1982) An elevator starts on the top floor of a 100-floor building and in its descent to the bottom (first) floor stops at least 40 floors, counting both the top and bottom floors as stops. Show that somewhere in its travel the elevator had to stop at two floors that were exactly 9, 10, or 19 floors apart.

9.33 (a) Show that if p is a prime of the form $4n + 1$, then there exists x such that

$$x^2 \equiv -1 \bmod p.$$

Hint: Use Wilson's theorem (Problem 7.6).

(b) Show that if p is a prime number of the form $4n + 1$, then p is expressible as the sum of two squares of integers.

9.34 (a) (Blichfeldt's lemma) Show that for any plane region R with area greater than n units, one of the translates of R contains at least $n + 1$ lattice points.

Hint: There exists a point (x, y) such that R contains $n + 1$ points of the form $(x + m, y + n)$, $m, n \in \mathbf{Z}$, or else R would have area at most n.

(b) Suppose that R is a plane region of area greater than 1. Show that R contains two points, (x_1, y_1) and (x_2, y_2), such that $x_1 - x_2$ and $y_1 - y_2$ are integers.

Hint: Since the area of the given region is greater than 1, we may translate it to contain two lattice points (x_1', y_1') and (x_2', y_2').

(c) (Minkowski's theorem) Let A be a convex set in the plane with the property that if (x, y) belongs to A then $(-x, -y)$ belongs to A. Show that if the area of A is greater than 4, then A contains a lattice point other than the origin.

9.35 Let A_1 be a two-row array of positive integers

$$
\begin{array}{cccc}
a_1 & a_2 & \dots & a_m \\
b_1 & b_2 & \dots & b_m
\end{array},
$$

where the a_i are distinct and written in increasing order. Let $c_1, \dots, c_n$ (also written in increasing order) be the list of all integers which occur in A_1, and let d_i, $1 \le i \le n$, be the number of occurrences of c_i in A_1. Let A_2 be the array

$$
\begin{array}{cccc}
c_1 & c_2 & \dots & c_n \\
d_1 & d_2 & \dots & d_n
\end{array}.
$$

For example, if A_1 is

$$
\begin{array}{ccccc}
1 & 2 & 5 & 7 & 8 \\
1 & 2 & 3 & 1 & 1
\end{array},
$$

then A_2 is

$$
\begin{array}{cccccc}
1 & 2 & 3 & 5 & 7 & 8 \\
4 & 2 & 1 & 1 & 1 & 1
\end{array}.
$$

Starting with any array A_1, array A_2 is created as specified above, and then the process is repeated to create A_3 from A_2, and so on. Show that the number of distinct arrays produced in this manner is always finite. In particular, show that the process always "goes into a loop" consisting of one, two, or three arrays.

Chapter 10

Two-Way Counting

> Straight up, straight down; two and two made four, and screw
> it around as much as you like, it didn't make three-and-a-half,
> and it didn't make four-and-a-half. It made *four*, period.

JOHN WAINWRIGHT
All on a Summer's Day, 1981

This chapter has to do with solutions in which one makes a count or an observation in two different ways and compares the results.

Example 10.1. Prove *Pascal's identity*:

$$\binom{n+1}{k} = \binom{n}{k} + \binom{n}{k-1},\tag{10.1}$$

for $1 \leq k \leq n$.

Solution: The left side of the equation (10.1) represents the number of k-element subsets of a set X of cardinality $n+1$. Let us see that the right side of (10.1) counts the same thing. Choose $x \in X$. Each k-subset of X either contains x or it does not. If it does, there are $\binom{n}{k-1}$ choices for the other elements of the subset. If it doesn't, all k elements must be selected from the remaining n elements of X, and there are $\binom{n}{k}$ choices. Altogether the number of k-subsets of an $(n+1)$-set is $\binom{n}{k-1} + \binom{n}{k}$. The two sides of (10.1) count the same thing and are therefore equal. ∎

Example 10.2. Prove the identity

$$\binom{n}{j}\binom{j}{k} = \binom{n}{k}\binom{n-k}{j-k}.\tag{10.2}$$

Solution: Let X be a set of n elements. Suppose that we want to select subsets Y and Z of X with $Z \subseteq Y \subseteq X$, $|Y| = j$, and $|Z| = k$. Clearly, this can be done in $\binom{n}{j}\binom{j}{k}$ ways by choosing Y and then choosing Z. But we can also choose Z first and then choose Y. As Y must contain all k elements of Z, we are free to choose $j - k$ elements from the $n - k$ elements of X not already chosen. This task can be performed in $\binom{n}{k}\binom{n-k}{j-k}$ ways. We have described two different counts for the same collection, and the identity (10.2) follows.

Note: We can think of the identity (10.2) as two ways of choosing, from n people, a committee of size k and a subcommittee of size j. Hence (10.2) is called the "subcommittee identity."   ∎

Example 10.3. Prove that

$$\binom{2n}{n} = \sum_{k=0}^{n}\binom{n}{k}^2.$$

Solution: As in Figure 10.1, the binomial coefficient $\binom{2n}{n}$ is the number of northeast paths which start at the southwest corner A of an $n \times n$ grid and stop at the northeast corner B. Such paths have length $2n$ and are determined by a sequence of n "easts" and n "norths" in some order. The number of such sequences is $\binom{2n}{n}$. The summation $\sum_{k=0}^{n}\binom{n}{k}^2$ counts the paths according to their intersection with the main diagonal of the grid. That is, the number of paths which cross the diagonal at the point $(k, n-k)$ is $\binom{n}{k}^2$, where $0 \le k \le n$.

Question: Is there a similar argument by which we can calculate

$$\sum_{k=0}^{n}\binom{n}{k}^3 ?$$

■

Problems

10.1 Prove that

$$\binom{m+n+1}{n+1} = \sum_{i=0}^{m}\binom{n+i}{n}.$$

10.2 Prove that

$$(-1)^k\binom{m-1}{k} = \sum_{i=0}^{k}\binom{m}{i}(-1)^i.$$

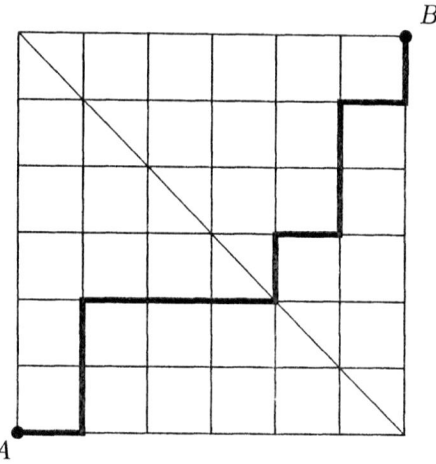

B

A

Figure 10.1: One of many northeast paths in a square grid.

10.3 Suppose that A_1, ..., A_{100} are subsets of a finite set S, each with $|A_i| > 2|S|/3$. Prove that there exists $x \in S$ with x contained in at least 67 of the A_i. Also, show that 67 is the best possible result.

10.4 Prove that in the definition of field (see Glossary), we can omit the requirement that '+' is a commutative operation.

10.5 (International Olympiad, 1977) In a finite sequence of real numbers the sum of any seven consecutive terms is negative and the sum of any eleven consecutive terms is positive. Determine the maximum number of terms in the sequence.

Solutions

10.1 The binomial coefficient $\binom{m+n+1}{n+1}$ is the number of $(n+1)$-subsets of the set $\{1, \ldots, m+n+1\}$. Suppose that the largest element in such a subset is $n+i+1$, where $0 \le i \le m$. The other n elements are chosen from among the elements $1, \ldots, n+i$; there are $\binom{n+i}{n}$ choices. Therefore the summation $\sum_{i=0}^{m} \binom{n+i}{n}$ counts all $(n+1)$-subsets.

10.2 We start with the algebraic identity

$$(1+x)^{m-1} = (1+x)^m (1 - x + x^2 - x^3 + \cdots). \tag{10.3}$$

Equating coefficients of x^k on the two sides of this equation yields the desired identity.

10.3 Suppose that there exists no element $x \in S$ contained in at least 67 of the A_i. Then each x is contained in at most 66 of the A_i. But then

$$|\{(x, i) : x \in A_i, 1 \leq i \leq 100\}| \leq 66 \cdot |S|,$$

while we are given that

$$|\{(x, i) : x \in A_i, 1 \leq i \leq 100\}| > \frac{2}{3} \cdot 100 \cdot |S|,$$

a contradiction.

To show that this result is best possible, let $S = \{1, \ldots, 100\}$ and take A_1 to be any subset of S with 67 elements. Let $A_1, A_2, \ldots, A_{100}$ be successive cyclic shifts of A_1. In this system of sets, each $x \in S$ is contained in exactly 67 of the A_i.

10.4 Let x and y be any elements of the field. We evaluate the expression

$$(1 + 1) \cdot (x + y)$$

in two different ways. First,

$$\begin{aligned}(1 + 1) \cdot (x + y) &= 1 \cdot (x + y) + 1 \cdot (x + y) \\ &= (x + y) + (x + y).\end{aligned}$$

Second,

$$\begin{aligned}(1 + 1) \cdot (x + y) &= (1 + 1) \cdot x + (1 + 1) \cdot y \\ &= (x + x) + (y + y).\end{aligned}$$

It follows that
$$x + y + x + y = x + x + y + y.$$

Canceling an x on the left and a y on the right, we obtain

$$x + y = y + x.$$

10.5 The maximum possible number of terms is sixteen. An example of a sixteen term sequence of the required type is

$$5, \ 5, \ -13, \ 5, \ 5, \ 5, \ -13, \ 5, \ 5, \ -13, \ 5, \ 5, \ 5, \ -13, \ 5, \ 5.$$

Now assume that $x_1, \ldots, x_{17}$ is such a sequence with seventeen terms. Let S be the sum of all possible sums of seven consecutive terms:

$$
\begin{aligned}
S \;=\; & (x_1 + x_2 + \cdots + x_7) \\
& + (x_2 + x_3 + \cdots + x_8) \\
& + \cdots \\
& + (x_{11} + x_{12} + \cdots + x_{17}).
\end{aligned}
$$

Because the sum of any seven consecutive terms is negative, S is negative. However, S may also be written as the sum of every possible sum of eleven consecutive terms:

$$
\begin{aligned}
S \;=\; & (x_1 + x_2 + \cdots + x_{11}) \\
& + (x_2 + x_3 + \cdots + x_{12}) \\
& + \cdots \\
& + (x_7 + x_8 + \cdots + x_{17}).
\end{aligned}
$$

Because the sum of any eleven consecutive terms is positive, S is positive. But we have said that S is both positive and negative. This contradiction shows that there is no such sequence of seventeen terms.

Additional Problems

10.6 Suppose that a_1, a_2, and a_3 are altitudes of a triangle and r is the radius of the triangle's inscribed circle. Show that

$$
\frac{1}{a_1} + \frac{1}{a_2} + \frac{1}{a_3} = \frac{1}{r}.
$$

10.7 Prove the identity

$$
\sum_{k=0}^{n} k \binom{n}{k} = n 2^{n-1}.
$$

Note: This problem also appears as Additional Problem 13.12.

10.8 Prove the identity

$$
\sum_{i=0}^{n} \binom{n+i}{n} 2^{-i} = 2^n.
$$

10.9 Prove the identity

$$\binom{2n-1}{n} = \sum_{k=0}^{n} \binom{n}{k}\binom{n-1}{k}.$$

10.10 Prove the identity

$$\sum_{i=0}^{n} \binom{n}{i}\binom{2n}{i} = \binom{3n}{n}.$$

10.11 Prove *Vandermonde's indentity:*

$$\binom{m+n}{n} = \sum_{k=0}^{n} \binom{n}{k}\binom{m}{n-k}.$$

10.12 Prove that

$$\sum_{d|n} \phi(d) = n,$$

where ϕ is Euler's phi-function (Glossary).

10.13 (*Math. Magazine*, Problem 924, January 1975; modified) If V is a set of k elements, how many n-tuples $(S_1, S_2, \ldots, S_n)$ exist with $S_1 \subseteq S_2 \subseteq \cdots \subseteq S_n \subseteq V$?

10.14 (*School Science and Mathematics*, Problem 4221, 1989; modified) Give a combinatorial interpretation and proof of

$$\sum_{i=0}^{\lfloor n/2 \rfloor} \binom{n}{i}\binom{n-i}{i} 2^{n-2i} = \binom{2n}{n}$$

where $n \geq 1$ and $\lfloor x \rfloor$ denotes the greatest integer less than or equal to x.

Chapter 11

Inclusion–Exclusion Principle

"If there is no way out, the best course of action is to find a way further in."

ORANIA PAPAZOGLOU
Pay McKenna, *Wicked, Loving Murder*, 1985

The inclusion–exclusion principle is a generalization of the Venn diagram rule,

$$|A \cup B| = |A| + |B| - |A \cap B|,$$

where A and B are finite sets. See Figure 11.1.

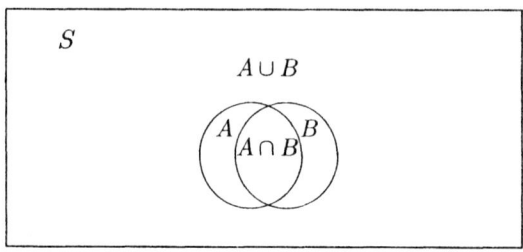

Figure 11.1: Venn diagram for two sets.

Theorem 11.1 (Inclusion–exclusion principle). Suppose that $A_1, \ldots,$

115

A_n are subsets of a finite set S. For $1 \leq k \leq n$, define

$$I_{n,k} = \sum_{1 \leq i_1 < \cdots < i_k \leq n} |A_{i_1} \cap \cdots \cap A_{i_k}|.$$

Then

$$|A_1 \cup \cdots \cup A_n| = \sum_{k=1}^{n} (-1)^{k+1} I_{n,k}. \tag{11.1}$$

Proof. The case $n = 1$ is trivial and the case $n = 2$ is established by Figure 11.1. Now assume that the formula holds for a certain value of n. Then

$$\begin{aligned}
|A_1 \cup \cdots \cup A_{n+1}| &= |A_1 \cup \cdots \cup A_n| + |A_{n+1}| - |(A_1 \cup \cdots \cup A_n) \cap A_{n+1}| \\
&= |A_1 \cup \cdots \cup A_n| + |A_{n+1}| \\
&\quad - |(A_1 \cap A_{n+1}) \cup \cdots \cup (A_n \cap A_{n+1})| \\
&= \sum_{k=1}^{n} (-1)^{k+1} I_{n,k} + |A_{n+1}| \\
&\quad - \sum_{k=1}^{n} (-1)^{k+1} \sum_{1 \leq i_1 < \cdots < i_k \leq n} |A_{i_1} \cap \cdots \cap A_{i_k} \cap A_{n+1}| \\
&= \sum_{k=1}^{n+1} (-1)^{k+1} I_{n+1,k}.
\end{aligned}$$

Therefore, by induction, the formula (11.1) holds for all $n \geq 1$. ∎

Example 11.1. A *derangement* of a set is permutation of the set with no fixed points. Let D_n be the number of derangements of the set $\{1, \ldots, n\}$.

(a) Find a formula for D_n.

(b) Let P_n be the probability that a random permutation of the set $\{1, \ldots, n\}$ is a derangement, i.e.,

$$P_n = \frac{D_n}{n!}.$$

Compute

$$\lim_{n \to \infty} P_n.$$

Solution: (a) For $1 \le i \le n$, let A_i be the collection of permutations of $\{1, \ldots, n\}$ which have i as a fixed point. Then

$$I_{n,k} = \binom{n}{k}(n-k)!,$$

as there are $\binom{n}{k}$ choices for which k elements are fixed and $(n-k)!$ permutations of the other elements. Therefore, by the formula (11.1),

$$|A_1 \cup \cdots \cup A_n| = \sum_{k=1}^{n}(-1)^k\binom{n}{k}(n-k)!.$$

Hence

$$D_n = n! - n!\sum_{k=1}^{n}\frac{(-1)^{k+1}}{k!} = n!\sum_{k=0}^{n}\frac{(-1)^k}{k!}. \qquad (11.2)$$

(b) From (a),

$$P_n = \sum_{k=0}^{n}\frac{(-1)^k}{k!}.$$

Hence

$$\lim_{n\to\infty} P_n = e^{-1} \doteq 0.37.$$

∎

Example 11.2. How many onto functions are there from the set $X = \{1, \ldots, m\}$ to the set $Y = \{1, \ldots, n\}$?

Solution: For $1 \le i \le n$, let A_i be the collection of functions whose ranges do not contain i. Then the intersection of k of the A_i has cardinality $(n-k)^m$, i.e., the number of functions from X to the nonexcluded $n-k$ elements of Y. Applying the inclusion–exclusion principle, we obtain an expression for the number of onto functions:

$$n^m - |A_1 \cup \cdots \cup A_n| = n^m - \sum_{k=1}^{n}(-1)^{k+1}\binom{n}{k}(n-k)^m$$

$$= \sum_{k=0}^{n}(-1)^k\binom{n}{k}(n-k)^m.$$

∎

Example 11.3. Euler's ϕ function is defined for all natural numbers by

$$\phi(n) = |\{1 \le x \le n : \gcd(x, n) = 1\}|.$$

Find a formula for $\phi(n)$.

Solution: Let the canonical factorization of n into prime powers be

$$n = \prod_{i=1}^{k} p_i^{\alpha_i}.$$

For $1 \leq i \leq k$, let

$$X_i = \{y : 1 \leq y \leq n \quad \text{and} \quad p_i \mid y\}.$$

Then

$$
\begin{aligned}
\phi(n) &= n - |X_1 \cup \cdots \cup X_k| \\
&= n - \left(\frac{n}{p_1} + \frac{n}{p_2} + \cdots\right) + \left(\frac{n}{p_1 p_2} + \frac{n}{p_1 p_3} + \cdots\right) - \cdots \\
&\quad + (-1)^k \frac{n}{p_1 p_1 \cdots p_k} \\
&= n\left(1 - \frac{1}{p_1}\right) \cdots \left(1 - \frac{1}{p_k}\right).
\end{aligned}
$$

■

Problems

11.1 (Putnam Competition, 1983) How many positive integers n are there such that n is an exact divisor of at least one of the numbers 10^{40}, 20^{30}?

11.2 (Putnam Competition, 1958) Show that the number of nonzero terms in the expansion of the nth order determinant having zeros in the main diagonal and ones elsewhere is

$$n! \left[1 - \frac{1}{1!} + \frac{1}{2!} - \frac{1}{3!} + \cdots + \frac{(-1)^n}{n!}\right].$$

11.3 (*Math. Magazine*, Problem 979, May 1976) Define $P(m, n)$ to be the number of permutations of the first n natural numbers for which m is the first number whose position is left unchanged. Clearly,

$$P(1, n) = (n - 1)!$$

for all n. Show that, for $m = 1, 2, \ldots, n - 1$,

$$P(m + 1, n) = P(m, n) - P(m, n - 1).$$

11.4 What is the expected number of fixed points in a permutation of $\{1, \ldots, n\}$?

Solutions

11.1 Let $\tau(n)$ be the number of divisors of the integer n. Then the number of divisors of 10^{40} or 20^{30} is

$$\tau\left(2^{40}5^{40}\right) + \tau\left(2^{60}5^{30}\right) - \tau\left(2^{40}5^{30}\right) = 41^2 + 61 \cdot 31 - 41 \cdot 31 = 2301.$$

11.2 The number of nonzero terms in the expansion of the determinant is clearly equal to the number of derangements of an n-element set. Thus the solution follows directly from the formula of Example 11.1.

11.3 For $1 \leq k \leq m - 1$, there are $(n - 1 - k)!$ permutations of $\{1, \ldots, n\}$ which fix m and also fix some given k-subset of $\{1, \ldots, m - 1\}$. Since the k-subset can be chosen in $\binom{m-1}{k}$ ways, it follows from Theorem 11.1 that

$$\begin{aligned} P(m, n) &= (n - 1)! - \sum_{k=1}^{m-1} (-1)^{k+1} \binom{m-1}{k}(n - 1 - k)! \\ &= \sum_{k=0}^{m-1} (-1)^k \binom{m-1}{k}(n - 1 - k)!. \end{aligned}$$

The recurrence relation is now proved easily:

$$\begin{aligned} P(m + 1, n) &= \sum_{k=0}^{m} (-1)^k \binom{m}{k}(n - 1 - k)! \\ &= \sum_{k=0}^{m} (-1)^k \left[\binom{m-1}{k} + \binom{m-1}{k-1}\right](n - 1 - k)! \\ &= P(m, n) + \sum_{k=0}^{m-1} (-1)^k \binom{m-1}{k}(n - 2 - k)! \\ &= P(m, n) - P(m, n - 1). \end{aligned}$$

11.4 The expected number of fixed points in a permutation of $\{1, \ldots, n\}$ is

$$E_n = \frac{1}{n!} \sum_{i=1}^{n} i \binom{n}{i} D_{n-i},$$

where D_n is given by the formula (11.2). Hence

$$E_n = \sum_{i=1}^{n} \frac{i}{i!(n - i)!} \sum_{j=0}^{n-i} (-1)^j \frac{(n - i)!}{j!}$$

$$= \sum_{i=0}^{n-1} \sum_{j=0}^{n-i-1} \frac{(-1)^j}{i!j!}$$

$$= \sum_{k=0}^{n-1} \sum_{i+j=k} \frac{(-1)^j}{i!j!}$$

$$= 1 + \sum_{k=1}^{n-1} \frac{1}{k!} \sum_{j=0}^{k} (-1)^j \binom{k}{j}$$

$$= 1 + \sum_{k=1}^{n-1} \frac{1}{k!} \cdot 0$$

$$= 1.$$

Note: We give a second proof that $E_n = 1$, using random variables. Given a random permutation of $\{1, 2, \ldots, n\}$ and $1 \leq i \leq n$, let $X_i = 1$ if i is a fixed point and $X_i = 0$ otherwise. Let $X = X_1 + \cdots + X_n$. Then $E_n = E(X)$, the expected value of X. By linearity of expectation,

$$E(X) = E(X_1 + \cdots + X_n) = E(X_1) + \cdots + E(X_n).$$

Since all the X_i clearly have the same expected value, namely, $1/n$, we have

$$E(X) = n \cdot \frac{1}{n} = 1.$$

Additional Problems

11.5 How many integers between 1 and 10^{100} are not perfect squares or perfect fifth powers?

11.6 How many positive integers are divisors of at least one of the numbers 12^{50}, 45^{20}, and 50^{100}?

11.7 How many sets are subsets of at least one of the sets $\{2, 4, 6, \ldots, 2000\}$, $\{3, 6, 9, \ldots, 3000\}$, and $\{5, 10, 15, \ldots, 5000\}$?

11.8 How many ways may n married couples sit around a table so that men and women sit in alternate seats and no person sits next to his spouse?

Note: This question is called the *Problème des ménages*. See [7].

11.9 (Bonferonni's inequalities) Suppose that S is a finite set, A_1, ..., $A_n \subseteq S$, and $1 \le m \le n$. Prove that

$$|A_1 \cup \cdots \cup A_n| \le \sum_{k=1}^{m} (-1)^{k+1} I_{n,k},$$

if m is odd, and the inequality is reversed if m is even.

11.10 (*Math. Magazine*, Problem 960, November 1975; modified)

(a) Suppose that a rectangle of size $a \times b$ is divided into ab unit squares by lines parallel to the sides of the rectangle. Show that a main diagonal of the rectangle passes through the interiors of exactly $a + b - \gcd(a, b)$ of the unit squares.

(b) Suppose that a box in $\mathbf{R}^d$ with dimensions a_1, a_2, ..., a_n is partitioned into $a_1 a_2 \ldots a_n$ unit d-dimensional cubes. Show that a main diagonal of the box passes through the interiors of exactly

$$\sum_i a_i - \sum_{i<j} \gcd(a_i, a_j) + \cdots + (-1)^{n+1} \gcd(a_1, a_2, \ldots, a_n)$$

of these cubes.

11.11 (*Math. Magazine*, Problem 1304, October 1988) For nonnegative integers k, n, establish the identity

$$\sum_{i \ge 0} (-1)^i \binom{n-k+1}{i} \binom{n-3i}{n-k} = \sum_{i \ge 0} \binom{n-k+1}{i} \binom{i}{k-i}.$$

(Here $\binom{n}{k} = 0$ when $k < 0$ or $k > n$.)

11.12 (*Math. Magazine*, Problem 1401, June 1992; modified) Let $f(n)$ be the number of permutations π of 1, 2, ..., $2n$ with the property that $|\pi(i+1) - \pi(i)| = n$ for some i, $1 \le i \le 2n - 1$. Show that

$$f(n) = \sum_{k=1}^{n} (-1)^{k-1} \binom{n}{k} 2^k (2n - k)!.$$

Chapter 12

Algebra of Polynomials

"If two and two plus x makes six, then to discover the unknown quantity is the simplest thing in the world."

MARY ROBERTS RINEHART
Miss Rachel Innes, *The Circular Staircase*, 1908

Let P be a polynomial of degree n with complex coefficients, i.e.,

$$P(x) = a_n x^n + a_{n-1} x^{n-1} + \cdots + a_0,$$

with all $a_i \in \mathbf{C}$ and $a_n \neq 0$. We say that a complex number r is a *root* of P if $P(r) = 0$. If $P(x) = a(x)(x - r)$ for some polynomial a, i.e., $x - r$ is a factor of P, then $P(r) = a(r)(r - r) = 0$, so that r is a root of P. The converse is also true, as we will demonstrate with the help of the following theorem, proved in most books on abstract algebra.

Theorem 12.1 (Division relation for polynomials). Suppose that f and g are polynomials with complex coefficients and positive degree. Then there exist unique polynomials a and b, also with complex coefficients, such that

$$f(x) = a(x)g(x) + b(x),$$

and either $b(x) = 0$ or $\deg b < \deg g$.

Suppose that r is a root of P. By Theorem 12.1, $P(x) = a(x)(x - r) + c$, where c is a constant. It follows that $0 = P(r) = a(r) \cdot 0 + c$, and therefore $c = 0$ and $P(x) = a(x)(x - r)$; hence $x - r$ is a factor of P.

The existence of roots of a polynomial is known as the "fundamental theorem of algebra."

Theorem 12.2 (Fundamental theorem of algebra). Let

$$P(x) = a_n x^n + a_{n-1} x^{n-1} + \cdots + a_0 \qquad (12.1)$$

be a polynomial of positive degree n with complex coefficients. Then P has a complex root.

Although the proof of the fundamental theorem is difficult, the following corollary is easy to prove by repeated use of the division argument above.

Corollary 12.3. The polynomial (12.1) has n roots counting multiplicity. Furthermore, if the roots are $r_1, \ldots, r_n$, then

$$P(x) = a_n(x - r_1) \ldots (x - r_n).$$

Example 12.1. Assume that a and b are different roots of $p(x) = x^3 + x - 1$. Prove that ab is a root of $q(x) = x^3 - x^2 - 1$.

Solution: Let the roots of p be a, b, and c. Then

$$x^3 + x - 1 = (x - a)(x - b)(x - c),$$

so that $abc = 1$. Hence

$$q(ab) = q\left(\frac{1}{c}\right) = \frac{1}{c^3} - \frac{1}{c^2} - 1 = -\frac{(c^3 + c - 1)}{c^3} = 0.$$

■

The example above illustrates the relationship between the coefficients of a polynomial and its roots. In general, given variables $x_1, \ldots, x_n$, the *elementary symmetric polynomials* are the polynomials

$$p_1 = x_1 + \cdots + x_n$$

$$p_2 = x_1 x_2 + x_1 x_3 + \cdots + x_{n-1} x_n$$

$$\vdots$$

$$p_n = x_1 \ldots x_n.$$

By direct calculation, we see that if

$$x^n + a_{n-1} x^{n-1} + \cdots + a_0 = (x - r_1)(x - r_2) \ldots (x - r_n),$$

then, for $i = 0, \ldots, n - 1$,

$$a_i = (-1)^{n-i} p_{n-i}, \qquad (12.2)$$

where the p_i are the elementary symmetric polynomials in the variables $r_1, \ldots, r_n$. The equation (12.2) displays the relationship between the coefficients of a polynomial and its roots.

Example 12.2. Assume that the roots of the polynomial

$$P(x) = x^4 + ax^3 + bx^2 + cx + d$$

are all real negative numbers. Prove that $a \geq 4d^{1/4}$.

Solution: Let the roots of P be r_1, r_2, r_3, r_4. Then $a = -r_1 - r_2 - r_3 - r_4$ and $d = r_1 r_2 r_3 r_4$. By the AM–GM inequality (Chapter 15), $a/4 \geq d^{1/4}$, and the result follows. ∎

Theorem 12.4 (Lagrange's interpolation formula). Let α_0, α_1, α_2, ..., α_n be distinct complex numbers and β_0, β_1, β_2, ..., β_n arbitrary complex numbers.

(a) The polynomial given by the formula

$$P(x) = \sum_{i=0}^{n} \beta_i \prod_{\substack{0 \leq j \leq n \\ j \neq i}} \frac{x - \alpha_j}{\alpha_i - \alpha_j}$$

has the property that $P(\alpha_i) = \beta_i$ for $i = 0, 1, \ldots, n$.

(b) There is exactly one polynomial P of degree at most n with the property specified in (a).

Proof. (a) The result is verified by evaluating P at α_0, α_1, α_2, ..., α_n.

(b) Suppose that there are two such polynomials of degree at most n, P and Q. Let $R(x) = P(x) - Q(x)$. Then R is of degree at most n and has $n + 1$ roots, namely, α_0, α_1, α_2, ..., α_n. Hence R is identically 0, and P and Q are the same polynomial. This establishes uniqueness. ∎

Example 12.3. A polynomial P of degree at most three assumes the values 1, 2, 3, 4 for $x = 1, 2, 3, 4$ (though not necessarily in that order). What is the largest possible value of $P(5)$?

Solution: By Lagrange's interpolation formula, we have

$$P(x) = P(1)\frac{(x-2)(x-3)(x-4)}{(1-2)(1-3)(1-4)} + P(2)\frac{(x-1)(x-3)(x-4)}{(2-1)(2-3)(2-4)}$$

$$+ P(3)\frac{(x-1)(x-2)(x-4)}{(3-1)(3-2)(3-4)} + P(4)\frac{(x-1)(x-2)(x-3)}{(4-1)(4-2)(4-3)}$$

so that

$$P(5) = -P(1) + 4P(2) - 6P(3) + 4P(4).$$

To maximize $P(5)$, we let $P(3) = 1$, $P(1) = 2$, $P(4) = 3$, and $P(2) = 4$, and obtain $P(5) = 20$. ∎

The following proposition is found in most books on abstract algebra and is needed for the forthcoming theorem on partial fractions decomposition.

Theorem 12.5. If two polynomials f and g have no common factor, then there exist polynomials a and b such that

$$a(x)f(x) + b(x)g(x) = 1.$$

Theorem 12.6 (Partial fractions decomposition). Suppose that

$$f(x) = \frac{g(x)}{(x - r_1)^{m_1} \ldots (x - r_n)^{m_n}},$$

where g is a polynomial. Then

$$f(x) = h(x) + \sum_{i=1}^{n} \sum_{j=1}^{m_i} \frac{c_{ij}}{(x - r_i)^j}$$

for some polynomial h and constants c_{ij}, $1 \le i \le n$, $1 \le j \le m_i$.

Note: The case of nonrepeated roots follows easily from Lagrange's interpolation formula. For suppose that

$$f(x) = \frac{g(x)}{(x - r_1) \ldots (x - r_n)},$$

where the r_i are distinct. Write

$$f(x) = h(x) + \frac{\overline{g}(x)}{(x - r_1) \ldots (x - r_n)},$$

where $\deg \overline{g}(x) < n$. By Lagrange's interpolation formula,

$$\overline{g}(x) = \sum_{i=1}^{n} \overline{g}(r_i) \prod_{\substack{0 \le j \le n \\ j \ne i}} \frac{x - r_j}{r_i - r_j},$$

so, dividing by $(x - r_1) \ldots (x - r_n)$, we obtain

$$f(x) = h(x) + \sum_{i=1}^{n} \frac{c_i}{x - r_i}$$

where the c_i are constants.

Proof. Since $(x - r_1)^{m_1}$ and $(x - r_2)^{m_2} \ldots (x - r_n)^{m_n}$ have no common factors, there are polynomials $a(x)$ and $b(x)$ for which

$$a(x)(x - r_1)^{m_1} + b(x)(x - r_2)^{m_2} \ldots (x - r_n)^{m_n} = 1.$$

Hence

$$\frac{g(x)}{(x - r_1)^{m_1} \ldots (x - r_n)^{m_n}} = \frac{a(x)g(x)}{(x - r_2)^{m_2} \ldots (x - r_n)^{m_n}} + \frac{b(x)g(x)}{(x - r_1)^{m_1}}$$

and the problem reduces to the case of just one root (with multiplicity). So assume that

$$f(x) = \frac{g(x)}{(x - r)^m}.$$

Using Theorem 12.1, we write $g(x) = \bar{g}(x)(x - r) + c$, where c is a constant. Hence

$$f(x) = \frac{\bar{g}(x)}{(x - r)^{m-1}} + \frac{c}{(x - r)^m},$$

and the result follows by induction. ∎

Remark: A similar argument shows that if f is a polynomial with real coefficients, then f has a partial fractions decomposition consisting of terms of the form

$$\frac{A}{(x - r)^k},$$

where A and r are real numbers, and/or terms of the form

$$\frac{Ax + B}{(x^2 + mx + n)^k},$$

where A, B, m, and n are real numbers and the quadratic $x^2 + mx + n$ is irreducible over **R**.

Example 12.4. Let $\frac{f(x)}{g(x)}$ be a rational function with $\deg f < \deg g$ and assume that $x - r$ is a nonrepeated linear factor of $g(x)$. Prove that the term in the partial fraction decomposition of $\frac{f(x)}{g(x)}$ corresponding to $x - r$ is $\frac{A}{x-r}$ where

$$A = \lim_{x \to r} \frac{(x - r)f(x)}{g(x)}.$$

Use this fact to find the partial fraction decomposition for

$$h(x) = \frac{2x^2 - 15x + 7}{(x - 1)(x + 2)(x - 3)}.$$

Solution: We have

$$\frac{f(x)}{g(x)} = \frac{A}{x-r} + \alpha(x)$$

where $\alpha(x)$ is continuous at $x = r$. Hence

$$\lim_{x \to r} \frac{(x-r)f(x)}{g(x)} = \lim_{x \to r} [A + \alpha(x)(x-r)] = A.$$

For the given $h(x)$,

$$\lim_{x \to 1} \frac{2x^2 - 15x + 7}{(x+2)(x-3)} = \frac{-6}{-6} = 1,$$

$$\lim_{x \to -2} \frac{2x^2 - 15x + 7}{(x-1)(x-3)} = \frac{45}{15} = 3,$$

$$\lim_{x \to 3} \frac{2x^2 - 15x + 7}{(x-1)(x+2)} = \frac{-20}{10} = -2.$$

Therefore

$$h(x) = \frac{1}{x-1} + \frac{3}{x+2} - \frac{2}{x-3}.$$

■

Problems

12.1 If the polynomial P given by

$$P(x) = a_n x^n + a_{n-1} x^{n-1} + \cdots + a_1 x + a_0$$

has distinct zeros $r_1, r_2, \ldots, r_n$, prove that

$$\frac{1}{r_1} + \frac{1}{r_2} + \cdots + \frac{1}{r_n} = -\frac{a_1}{a_0}.$$

Assume that no $r_i = 0$.

12.2 Show that

$$\sum_{j=0}^{n} \binom{n}{j}(-1)^j j^m = \begin{cases} 0 & 0 \le m \le n-1, \\ (-1)^n n! & m = n. \end{cases}$$

12.3 Suppose that k is a positive integer and P is a polynomial in one variable of degree n. Prove that if

$$Q_k(x) = \sum_{j=0}^{k} (-1)^{k-j} \binom{k}{j} P(x+j),$$

then

(a) $Q_k(x)$ is a polynomial of degree $n - k$ if $n \geq k$, and

(b) $Q_k(x)$ is identically zero if $n < k$.

12.4 Consider the sequence of polynomials

$$\binom{x}{0} = 1, \quad \binom{x}{1} = x, \quad \binom{x}{2} = \frac{x(x-1)}{2}, \quad \binom{x}{3} = \frac{x(x-1)(x-2)}{3!}, \ldots$$

(a) Prove that any polynomial P of degree n can be expressed uniquely in the form

$$P(x) = a_0\binom{x}{0} + a_1\binom{x}{1} + a_2\binom{x}{2} + a_3\binom{x}{3} + \cdots + a_n\binom{x}{n}.$$

(b) In (a), prove that the coefficients $a_0, a_1, a_2, \ldots, a_n$ are integral if and only if $P(j)$ is integral for all integers j.

12.5 Suppose that P is a polynomial in one variable with the property that $P(z)$ is real when and only when z is real. Prove that P must be a polynomial of first degree with real coefficients.

12.6 Prove that if the quadratics $ax^2 + bx + c$ and $px^2 + qx + r$ have a common zero then $(ar - cp)^2 = (aq - bp)(br - cq)$.

12.7 Prove that two polynomials P and Q with $P^2 - Q^3 = 1$ must be constants.

12.8 Let $P(x)$ be a polynomial of degree $n \geq 1$ with distinct roots $r_1, \ldots, r_n$. Show that for any ξ with $P'(\xi) \neq 0$ there exists i such that

$$|\xi - r_i| \leq n|P(\xi)/P'(\xi)|.$$

Solutions

12.1 We are given that $P(x) = a_n(x - r_1)(x - r_2)\ldots(x - r_n)$. Therefore

$$a_0 = a_n(-1)^n r_1 r_2 \ldots r_n$$

and

$$a_1 = a_n \sum_{i=1}^{n}(-1)^{n-1}\frac{r_1 \ldots r_n}{r_i}.$$

Dividing the second equation by the first we obtain the desired result.

12.2 Recall Example 11.2, where we showed that the number of onto functions from $X = \{1, \ldots, m\}$ to $Y = \{1, \ldots, n\}$ is

$$\sum_{j=0}^{n}(-1)^j \binom{n}{j}(n-j)^m.$$

As such, this expression equals 0 if $0 \le m \le n-1$ and $n!$ if $m = n$. The result now follows upon replacing j with $n-j$.

12.3 From the previous solution, the number of surjective maps from a set of size i to a set of size k is $\sum_{j=0}^{k}(-1)^{k-j}\binom{k}{j}j^i$, and hence we have

$$\sum_{j=0}^{k}(-1)^{k-j}\binom{k}{j}j^i = 0$$

for $i < k$. Therefore, for $m < k$,

$$\sum_{j=0}^{k}(-1)^{k-j}\binom{k}{j}(x+j)^m = \sum_{j=0}^{k}(-1)^{k-j}\binom{k}{j}\sum_{i=0}^{m}\binom{m}{i}x^{m-i}j^i$$

$$= \sum_{i=0}^{m}\binom{m}{i}x^{m-i}\sum_{j=0}^{k}(-1)^{k-j}\binom{k}{j}j^i$$

$$= 0.$$

This proves (b).

The inner summation above is 0 unless $i \ge k$. Hence, if $n > k$, then we have a nonzero summand when $m = n$ and $i = k$, yielding a polynomial of degree $n - k$. This proves (a).

12.4 (a) Letting $x = 0, \ldots, n$, we obtain the following equations:

$$P(0) = a_0$$
$$P(1) = a_0 + a_1$$
$$P(2) = a_0 + 2a_1 + a_2$$
$$P(3) = a_0 + 3a_1 + 3a_1 + a_3$$
$$P(4) = a_0 + \binom{4}{1}a_1 + \binom{4}{2}a_2 + \binom{4}{3}a_3 + a_4$$
$$\vdots$$

$$P(n-1) \quad = \quad a_0 + \binom{n-1}{1}a_1 + \cdots + \binom{n-1}{n-2}a_{n-2} + a_{n-1}$$

$$P(n) \quad = \quad a_0 + \binom{n}{1}a_1 + \cdots + \binom{n}{n-1}a_{n-1} + a_n.$$

This system is clearly solvable for $a_0, a_1, \ldots, a_n$, so $P(x)$ has an expression as required.

To show uniqueness, we note that if $P(x) = \sum_{k=0}^{n} a_k\binom{x}{k}$ and $P(x) = \sum_{k=0}^{n} b_k\binom{x}{k}$, then evaluating at $x = 0$ gives $a_0 = b_0$. Therefore

$$\sum_{k=1}^{n} a_k\binom{x}{k} = \sum_{k=1}^{n} b_k\binom{x}{k}.$$

Evaluating at $x = 1$ gives $a_1 = b_1$. Continuing, if

$$\sum_{k=j}^{n} a_k\binom{x}{k} = \sum_{k=j}^{n} b_k\binom{x}{k},$$

evaluating at $x = j$ gives $a_j = b_j$, etc. Thus we have proved part (a).

(b) This follows easily from the above system.

12.5 Assume that $\deg P = n$. If $P(z) = a(z - z_1)(z - z_2)\ldots(z - z_n)$, $a \neq 0$, then each z_i must be real. Choose a real z such that $P(z) \neq 0$. Then $a = P(z)/(z - z_1)\ldots(z - z_n)$, and a is real. Therefore P has all real coefficients.

Suppose that $n > 1$, If n is even then we can choose $y \in \mathbf{R}$ such that $y \notin f(\mathbf{R})$. But $f(z) = y$ has a solution in $\mathbf{C}$. This is a contradiction. If n is odd, $n > 1$, we can choose a real y such that $|f^{-1}(y) \cap \mathbf{R}| = 1$. Clearly there exists $z \in \mathbf{C} - \mathbf{R}$ such that $f(z) = y$. Therefore $n = 1$.

12.6 If $x = 0$ is a common root, then the result follows easily. If $ax^2 + bx + c$ and $px^2 + qx + r$ have a common nonzero root, then the system of equations

$$ax^3 + bx^2 + cx \quad = \quad 0$$

$$ax^2 + bx + c \quad = \quad 0$$

$$px^3 + qx^2 + rx \quad = \quad 0$$

$$px^2 + qx + r \quad = \quad 0$$

has a nontrivial solution. Therefore,

$$0 = \begin{vmatrix} a & b & c & 0 \\ 0 & a & b & c \\ p & q & r & 0 \\ 0 & p & q & r \end{vmatrix} = (ar - cp)^2 - (br - cq)(aq - bp).$$

Note: The above determinant is called the *resultant* of the two given polynomials. See [3].

Resultants can be used to determine whether a given polynomial f has a repeated root. It is easy to show that f has a repeated root if and only if f and f' have a common root, and this criterion can be checked as above. For example, let $f(x) = ax^2 + bx + c$ and hence $f'(x) = 2ax + b$. The resultant of f and f' is

$$0 = \begin{vmatrix} a & b & c & 0 \\ 0 & a & b & c \\ 0 & 2a & b & 0 \\ 0 & 0 & 2a & b \end{vmatrix} = -a^2(b^2 - 4ac).$$

(We have departed from the usual definition of resultant. Again, see [3].) The quantity $b^2 - 4ac$ is sometimes called the *discriminant* of f. If the discriminant is 0, then f has a repeated root.

12.7 Suppose that there are nonconstant polynomials $P(x)$ and $Q(x)$ with $P^2(x) - Q^3(x) = 1$. We note that $\deg P > \deg Q$. Furthermore, no zero of P is a zero of Q. Differentiating, we obtain the relation

$$P(x)P'(x) = 3Q^2(x)Q'(x).$$

By the fundamental theorem of algebra, all the linear factors of $P(x)$ occur on the right side of this relation. But as they are not factors of $Q(x)$, they are factors of $Q'(x)$ (with the same or greater multiplicity). Therefore, $\deg P \le \deg Q' < \deg Q$. But this inequality contradicts the previous inequality on the degrees of P and Q.

12.8 Let $P(x) = a_n(x - r_1)\ldots(x - r_n)$. Then

$$\ln|P(x)| = \ln|a_n| + \sum_{i=1}^{n} \ln|x - r_i|$$

and

$$\frac{P'(\xi)}{P(\xi)} = \sum_{i=1}^{n} \frac{1}{\xi - r_i}$$

$$\leq \sum_{i=1}^{n} \frac{1}{|\xi - r_i|}$$

$$\leq \frac{n}{|\xi - r_i|},$$

where we have taken the maximum term in the sum. The result follows directly.

Additional Problems

12.9 The *complex conjugate* of a complex number $z = a + bi$ is $\bar{z} = a - bi$.

(a) Show that if z is a root of of polynomial P with real coefficients, then $\bar{z}$ is a root of P.

(b) Show that if f is a polynomial with real coefficients, then the irreducible (over $\mathbf{R}$) factors of f are of degree one and/or two.

12.10 (*The Pentagon*, Problem 401, Fall 1987; modified) Show that if $b^2 < 3ac$ for real numbers a, b, and c, then the equation $x^3 + ax^2 + bx + c = 0$ has one real root and two non-real roots.

12.11 (*The Pentagon*, Problem 417, Fall 1988) Find r and s such that $rx^{15} - sx^{14} + 1$ is divisible by $x^2 - x - 1$.

12.12 (U. S. A. Olympiad, 1977) If a and b are two of the roots of $x^4 + x^3 - 1 = 0$, prove that ab is a root of $x^6 + x^4 + x^3 - x^2 - 1 = 0$.

12.13 (U. S. A. Olympiad, 1976; modified) If $P(x)$, $Q(x)$, $R(x)$, and $S(x)$ are polynomials for which

$$P(x^5) + xQ(x^5) + x^2R(x^5) = (x^4 + x^3 + x^2 + x + 1)S(x),$$

prove that $x - 1$ is a factor of $P(x)$.

12.14 Let $f(x)$ be a polynomial of degree n such that

$$f(k) = \frac{1}{k}$$

for each $k = 1, 2, \ldots, n + 1$. What is the value of $f(n + 2)$?

12.15 Prove that if n is a positive integer greater than 1, then the product of two polynomials of degree n with nonnegative coefficients cannot equal $x^{2n} + x^{2n-1} + \cdots + x + 1$.

Hint: Recall Example 5.1 !

Note: The irreducible (over **R**) factors of the polynomial

$$f_n(x) = x^n + x^{n-1} + x^{n-2} + \cdots + 1$$

are the *cyclotomic polynomials* $\Phi_d(x)$, where $d|n$. See [9].

12.16 Let $p(x)$ be a polynomial with real coefficients of degree $n \geq 2$ whose first three terms are $x^n + 2x^{n-1} + 2x^{n-2}$. Show that $p(x)$ cannot have n distinct real roots.

12.17 (Putnam Competition, 1963) For what integer a does $x^2 - x + a$ divide $x^{13} + x + 90$?

12.18 Let a, b, c be the roots of $x^3 - x^2 - x - 1$. Show that a, b, c are distinct and that

$$\frac{a^{1000} - b^{1000}}{a - b} + \frac{b^{1000} - c^{1000}}{b - c} + \frac{c^{1000} - a^{1000}}{c - a}$$

is an integer.

12.19 (Putnam Competition, 1971) Determine all polynomials $P(x)$ such that $P(x^2 + 1) = (P(x))^2 + 1$ and $P(0) = 0$.

12.20 Let S be the set of positive integers that have no 9 in their decimal expansions. Prove that $\sum_{n \in S} 1/n \leq 80$.

12.21 Suppose that the polynomial f is symmetric in the n variables x_1, $\ldots$, x_n. Prove that f is a polynomial in the elementary symmetric polynomials p_1, $\ldots$, p_n.

Chapter 13

Recurrence Relations and Generating Functions

"And now let us examine this matter of X."

AGATHA CHRISTIE
Hercule Poirot, *Curtain*, 1975

A *recurrence relation* is a rule for calculating the values of a function from previously known values. Recall Example 1.1, in which we found a recurrence relation for $f(n)$, the number of regions into which the plane is divided by n lines. The relation is:

$$f(n) = f(n-1) + n,$$

for all $n \geq 2$. This formula, together with the *initial value* $f(1) = 2$, allows us to compute any desired value of $f(n)$. (We also found an explicit formula for $f(n)$ in Chapter 1 but we are not concerned with that here.)

We say that a sequence $\{a_n\}$ is *defined recursively* if

$$a_n = f(a_{n-1}, a_{n-2}, \ldots, a_{n-k})$$

for $n \geq k+1$, where f is a function, and initial values a_1, ..., a_k are specified.

The above function $f(n)$ is defined recursively according to this definition.

As another example, recall that the Fibonacci numbers f_n are defined by the recurrence relation

$$f_n = f_{n-1} + f_{n-2},$$

135

for $n \geq 2$, where $f_0 = 0$ and $f_1 = 1$.

We say that a sequence $\{a_n\}$ satisfies a *linear recurrence relation with constant coefficients* if

$$a_n = c_1 a_{n-1} + c_2 a_{n-2} + \cdots + c_k a_{n-k}$$

for some k, all $n \geq k$, and constants $c_1, \ldots, c_k$.

The Fibonacci numbers satisfy a linear recurrence relation with constant coefficients. The function $f(n)$ above, in fact, does not.

An algebraic expression called a generating function is very useful for working on many kinds of sequences, including those satisfying linear recurrence relations with constant coefficients. There are two basic kinds of generating functions.

The *ordinary generating function* for the sequence $\{a_0, a_1, a_2, \ldots\}$ is given by

$$f(x) \;=\; \sum_{n=0}^{\infty} a_n x^n.$$

The *exponential generating function* for the same sequence is given by

$$g(x) \;=\; \sum_{n=0}^{\infty} \frac{a_n}{n!} x^n.$$

Example 13.1. Find the ordinary generating function for the Fibonacci sequence $\{f_0, f_1, f_2, \ldots\}$.

Solution: Let $f(x) = \sum_{n=0}^{\infty} f_n x^n$. Then

$$f(x) \;=\; x + x^2 + 2x^3 + 3x^4 + 5x^5 + \cdots$$
$$x f(x) \;=\; x^2 + x^3 + 2x^4 + 3x^5 + 5x^6 + \cdots$$
$$x^2 f(x) \;=\; x^3 + x^4 + 2x^5 + 3x^6 + 5x^7 + \cdots.$$

Therefore

$$f(x) - x f(x) - x^2 f(x) = x$$

and

$$f(x) = \frac{x}{1 - x - x^2}.$$

■

Notice that ordinary generating function for the Fibonacci sequence is a rational function. This situation is explained by the forthcoming Theorem 13.1.

Example 13.2. Evaluate

$$\sum_{n=1}^{\infty} f_n n 2^{-n}.$$

Solution: Let $f(x)$ be the generating function for the Fibonacci numbers determined in the previous example. Then

$$f'(x) = \sum_{n=1}^{\infty} f_n n x^{n-1},$$

$$x f'(x) = \sum_{n=1}^{\infty} f_n n x^n,$$

$$\frac{1}{2} f'\left(\frac{1}{2}\right) = \sum_{n=1}^{\infty} f_n n 2^{-n}$$

so that

$$\sum_{n=1}^{\infty} f_n n 2^{-n} = \frac{1}{2} \frac{d}{dx}\left(\frac{x}{1 - x - x^2}\right)\Big|_{x=1/2}$$

$$= 10.$$

■

A useful formula, needed in the proof of the following theorem, is

$$\frac{1}{(1 - x)^k} = \sum_{m=0}^{\infty} \binom{m + k - 1}{k - 1} x^m, \qquad (13.1)$$

for positive integers k. This formula can be proved by successively differentiating the power series for $(1 - x)^{-1}$. While we are not concerned about issues of convergence here, it is easy to show that the series (13.1) converges for $|x| < 1$.

Theorem 13.1. Let $\{a_n\}$ be a sequence of real or complex numbers and let $c_1, \ldots, c_k$ be real numbers. Then the following three statements are equivalent.
(1) The sequence $\{a_n\}$ satisfies a linear recurrence relation with constant coefficients $c_1, \ldots, c_k$, i.e.,

$$a_n = \sum_{i=1}^{k} c_i a_{n-i}$$

for $n \geq k$.

(2) The sequence $\{a_n\}$ has a rational ordinary generating function of the form

$$\frac{g(x)}{1 - \sum_{i=1}^{k} c_i x^i},$$

where g is a polynomial of degree at most $k - 1$.

(3) If

$$1 - \sum_{i=1}^{k} c_i x^i = (1 - r_1 x)(1 - r_2 x) \ldots (1 - r_k x),$$

with the r_i distinct, then the terms of $\{a_n\}$ are given by the formula

$$a_n = \alpha_1 r_1^n + \cdots + \alpha_k r_k^n,$$

for all $n \geq 0$, and arbitrary constants $\alpha_1, \ldots, \alpha_k$.

More generally, if

$$1 - \sum_{i=1}^{k} c_i x^i = (1 - r_1 x)^{m_1} (1 - r_2 x)^{m_2} \ldots (1 - r_i x)^{m_i},$$

where the roots $r_1, \ldots, r_i$ occur with multiplicities $m_1, \ldots, m_i$, then

$$a_n = p_1(n) r_1^n + \cdots + p_i(n) r_i^n$$

for all $n \geq 0$ and arbitrary polynomials $p_1, \ldots, p_i$, where $\deg p_i < m_i$.

Proof. (1) $\Rightarrow$ (2) Assume that $\{a_n\}$ satisfies a recurrence relation of the type specified in (1) and let $f(x)$ be the ordinary generating function for $\{a_n\}$. Then

$$
\begin{aligned}
f(x) &= \sum_{n=0}^{\infty} a_n x^n \\
&= \sum_{n=0}^{k-1} a_n x^n + \sum_{n=k}^{\infty} a_n x^n \\
&= \sum_{n=0}^{k-1} a_n x^n + \sum_{n=k}^{\infty} \sum_{i=1}^{k} c_i a_{n-i} x^n \\
&= \sum_{n=0}^{k-1} a_n x^n + \sum_{i=1}^{k} c_i \sum_{n=k}^{\infty} a_{n-i} x^n \\
&= \sum_{n=0}^{k-1} a_n x^n + \sum_{i=1}^{k} c_i x^i \sum_{n=k-i}^{\infty} a_n x^n
\end{aligned}
$$

$$= \sum_{n=0}^{k-1} a_n x^n + \sum_{i=1}^{k} c_i x^i \left[f(x) - \sum_{n=0}^{k-i-1} a_n x^n \right].$$

Hence

$$f(x) \left(1 - \sum_{i=1}^{k} c_i x^i \right) = \sum_{n=0}^{k-1} a_n x^n - \sum_{i=1}^{k} c_i x^i \sum_{n=0}^{k-i-1} a_n x^n,$$

so that

$$f(x) = \frac{g(x)}{1 - \sum_{i=1}^{k} c_i x^i},$$

where $\deg g \leq k - 1$.

(2) $\Rightarrow$ (3) First consider the case where

$$1 - \sum_{i=1}^{k} c_i x^i = (1 - r_1 x) \ldots (1 - r_k x),$$

and the r_i are distinct. Expanding by partial fractions, we obtain, for certain constants $\alpha_1, \ldots, \alpha_k$,

$$
\begin{aligned}
f(x) &= \frac{\alpha_1}{1 - r_1 x} + \cdots + \frac{\alpha_k}{1 - r_k x} \\
&= \alpha_1 \sum_{n=0}^{\infty} r_1^n x^n + \cdots + \alpha_k \sum_{n=0}^{\infty} r_k^n x^n \\
&= \sum_{n=0}^{\infty} (\alpha_1 r_1^n + \cdots + \alpha_k r_k^n) x^n.
\end{aligned}
$$

Since this is just another formula for the ordinary generating function for $\{a_n\}$, we have

$$a_n = \alpha_1 r_1^n + \cdots + \alpha_k r_k^n$$

for $n \geq 0$.

More generally, now assume that $1 - \sum_{i=1}^{k} c_i x^i$ has repeated roots. Suppose that r is a root with multiplicity k. Then, in the partial fraction decomposition of

$$\frac{g(x)}{1 - \sum_{i=1}^{k} c_i x^i},$$

we have terms

$$\frac{\beta_1}{(x - r)}, \frac{\beta_2}{(x - r)^2}, \ldots, \frac{\beta_k}{(x - r)^k},$$

where $\beta_1, \beta_2, \ldots, \beta_k$ are constants. By the formula (13.1), it follows that the contribution to x^i in the power series for these fractions is $P(n)x^n$, where P is a polynomial of degree less than k.

Clearly, the implications proved above are reversible.

Note: The initial values $a_0, \ldots, a_{k-1}$ of the recurrence satisfy the equations

$$a_0 = \alpha_1 + \cdots + \alpha_k$$

$$a_1 = \alpha_1 r_1 + \cdots + \alpha_k r_k$$

$$\vdots$$

$$a_{k-1} = \alpha_1 r_1^{k-1} + \cdots + \alpha_k r_k^{k-1},$$

and the determinant of this system of equations is

$$\begin{vmatrix} 1 & 1 & \cdots & 1 \\ r_1 & r_2 & \cdots & r_k \\ r_1^2 & r_2^2 & \cdots & r_k^2 \\ \vdots & \vdots & \vdots & \vdots \\ r_1^{k-1} & r_2^{k-1} & \cdots & r_k^{k-1} \end{vmatrix},$$

i.e., the Vandermonde determinant of Problem 4.13. This determinant equals

$$\prod_{1 \le i < j \le k} (r_j - r_i)$$

and hence is nonzero. Therefore, given the initial values $a_0, \ldots, a_{k-1}$, the values of $\alpha_1, \ldots, \alpha_k$ are uniquely determined. ∎

When factoring a polynomial, it is easier to find factors of the type $x - r$ than of the type $1 - rx$. Therefore, supposing

$$1 - \sum_{i=1}^{k} c_i x^i = (1 - r_1 x) \ldots (1 - r_k x),$$

we make the change of variables $x = y^{-1}$ and obtain the relation

$$1 - \sum_{i=1}^{k} c_i y^{-i} = (1 - r_1 y^{-1}) \ldots (1 - r_k y^{-1})$$

or

$$y^k - \sum_{i=1}^{k} c_i y^{k-i} = (y - r_1) \ldots (y - r_k). \tag{13.2}$$

The expression on the left of the equation (13.1) is called the *characteristic polynomial* of the sequence $\{a_n\}$.

Example 13.3. Find an explicit formula for the nth Fibonacci number f_n.

Solution: Since the recurrence relation for the Fibonacci sequence is

$$f_n = f_{n-1} + f_{n-2},$$

its characteristic polynomial is

$$x^2 - x - 1,$$

whose roots are

$$r_{1,2} = \frac{1 \pm \sqrt{5}}{2}.$$

By Theorem 13.1,

$$f_n = \alpha_1 r_1^n + \alpha_2 r_2^n,$$

where α_1 and α_2 are constants.

We use the initial values, $f_0 = 0$ and $f_1 = 1$, to solve for α_1 and α_2:

$$0 = \alpha_1 + \alpha_2$$

$$1 = \alpha_1 \left(\frac{1 + \sqrt{5}}{2} \right) + \alpha_2 \left(\frac{1 - \sqrt{5}}{2} \right).$$

We find that

$$\alpha_1 = \frac{1}{\sqrt{5}}$$

and

$$\alpha_2 = \frac{-1}{\sqrt{5}}.$$

Therefore

$$f_n = \frac{1}{\sqrt{5}} \left(\frac{1 + \sqrt{5}}{2} \right)^n - \frac{1}{\sqrt{5}} \left(\frac{1 - \sqrt{5}}{2} \right)^n$$

for $n \geq 0$. ∎

Example 13.4. Suppose that

$$a_n = 6a_{n-1} - 12a_{n-2} + 8a_{n-3},$$

for $n \geq 3$, and $a_0 = 0$, $a_1 = 2$, $a_2 = 64$. Find an explicit formula for a_n.

Solution: The characteristic polynomial of $\{a_n\}$ is

$$x^3 - 6x^2 + 12x - 8,$$

which has 2 as a root of multiplicity three. Hence

$$a_n = \alpha_1 2^n + \alpha_2 n 2^n + \alpha_3 n^2 2^n,$$

for certain constants α_1, α_2, α_3. Using the initial values, we find that $\alpha_1 = 0$, $\alpha_2 = -6$, and $\alpha_3 = 7$. Therefore,

$$a_n = -6n2^n + 7n^2 2^n,$$

for $n \geq 0$.   ∎

Example 13.5. Solve the differential equation

$$y'' - 5y' + 6y = 0$$

with initial conditions $y(0) = 0$ and $y'(0) = 1$.

Solution: We use an exponential generating function. Assume that

$$y = \sum_{n=0}^{\infty} a_n \frac{x^n}{n!}.$$

Then

$$y' = \sum_{n=1}^{\infty} a_n n \frac{x^{n-1}}{n!} = \sum_{n=1}^{\infty} a_n \frac{x^{n-1}}{(n-1)!} = \sum_{n=0}^{\infty} a_{n+1} \frac{x^n}{n!}$$

and

$$y'' = \sum_{n=1}^{\infty} a_{n+1} n \frac{x^{n-1}}{n!} = \sum_{n=1}^{\infty} a_{n+1} \frac{x^{n-1}}{(n-1)!} = \sum_{n=0}^{\infty} a_{n+2} \frac{x^n}{n!}.$$

Hence

$$\sum_{n=0}^{\infty} a_{n+2} \frac{x^n}{n!} - 5 \sum_{n=0}^{\infty} a_{n+1} \frac{x^n}{n!} + 6 \sum_{n=0}^{\infty} a_n \frac{x^n}{n!} = 0.$$

Therefore

$$\sum_{n=0}^{\infty} [a_{n+2} - 5a_{n+1} + 6a_n] \frac{x^n}{n!} = 0$$

and hence

$$a_{n+2} = 5a_{n+1} - 6a_n,$$

for $n \geq 0$. The characteristic polynomial of this recurrence relation is

$$y^2 - 5y + 6,$$

with roots

$$y = 2, \ 3.$$

Therefore

$$a_n = \alpha_1 2^n + \alpha_2 3^n$$

for $n \geq 0$, and hence

$$y = \sum_{n=0}^{\infty} \left[\frac{\alpha_1 2^n + \alpha_2 3^n}{n!} \right] x^n$$

$$= \alpha_1 e^{2x} + \alpha_2 e^{3x}.$$

The initial conditions give us two equations:

$$\alpha_1 + \alpha_2 = 0$$

$$2\alpha_1 + 3\alpha_2 = 1,$$

which yield $\alpha_1 = -1$, $\alpha_2 = 1$. Therefore

$$y = -e^{2x} + e^{3x}.$$

∎

The power series technique above applies to any linear homogeneous differential equation with constant coefficients. Suppose that

$$y^{(n)} = c_1 y^{(n-1)} + c_2 y^{(n-2)} + \cdots + c_k y^{(n-k)}.$$

The related recurrence relation for this equation is

$$a_n = c_1 a_{n-1} + \cdots + c_k a_{n-k}.$$

If the characteristic polynomial of this relation has distinct roots $r_1, \ldots, r_k$, then the complete solution to the differential equation is

$$y = \alpha_1 e^{r_1 x} + \cdots + \alpha_k e^{r_k x},$$

where the constants $\alpha_1, \ldots, \alpha_k$ are determined by the initial conditions.

Problems

13.1 Evaluate

$$\sum_{n=0}^{\infty} (n+1)^2 3^{-n}.$$

13.2 For the Fibonacci numbers $f_0 = 0$, $f_1 = 1$, $f_n = f_{n-1} + f_{n-2}$ for $n \geq 2$, determine constants a, b, c such that

$$f_n^2 = a f_{n-1}^2 + b f_{n-2}^2 + c f_{n-3}^2$$

for all $n \geq 3$.

13.3 Evaluate
$$\sum_{n=1}^{\infty} \frac{(-1)^{n+1}}{3n-2}.$$

13.4 Evaluate
$$\sum_{n=1}^{\infty} \frac{n^3}{n!}.$$

13.5 Calculate the remainder when $[(2+\sqrt{3})^{2^{1000}}]$ is divided by 17.

13.6 Show that
$$\prod_{k=1}^{n-1} \sin\left(\frac{\pi k}{n}\right) = n2^{1-n}.$$

13.7 Show that
$$\int_0^1 x^x \, dx = \sum_{k=1}^{\infty} (-1)^{k+1} k^{-k}.$$

Solutions

13.1 Since $\sum_{n=0}^{\infty} x^n = \frac{1}{1-x}$, we have $\sum_{n=0}^{\infty} x^{n+1} = \frac{x}{1-x}$ and
$$\sum_{n=0}^{\infty} (n+1)x^n = \frac{1(1-x) - x(-1)}{(1-x)^2} = \frac{1}{(1-x)^2}.$$

Hence
$$\sum_{n=0}^{\infty} (n+1)x^{n+1} = \frac{x}{(1-x)^2}$$

and
$$\sum_{n=0}^{\infty} (n+1)^2 x^n = \frac{1(1-x)^2 - 2x(1-x)(-1)}{(1-x)^4} = \frac{1+x}{(1-x)^3}.$$

Therefore
$$\sum_{n=0}^{\infty} (n+1)^2 3^{-n} = \left. \frac{1+x}{(1-x)^3} \right]_{1/3} = \frac{9}{2}.$$

13.2 Letting $n = 3$, 4, 5 yields the system
$$a + b = 4$$
$$4a + b + c = 9$$
$$9a + 4b + c = 25,$$

which has solution $a = 2$, $b = 2$, $c = -1$. Indeed, for all $n \geq 3$,

$$
\begin{aligned}
2f_{n-1}^2 + 2f_{n-2}^2 - f_{n-3}^2 &= f_{n-1}^2 + 2f_{n-1}f_{n-2} + \\
&\quad f_{n-2}^2 + f_{n-1}^2 - 2f_{n-1}f_{n-2} + f_{n-2}^2 - f_{n-3}^2 \\
&= (f_{n-1} + f_{n-2})^2 + (f_{n-1} - f_{n-2})^2 - f_{n-3}^2 \\
&= f_n^2 + f_{n-3}^2 - f_{n-3}^2 = f_n^2.
\end{aligned}
$$

13.3 The series $\sum_{i=0}^{\infty}(-1)^i x^{3i}$ converges uniformly to $\frac{1}{1+x^3}$ for $|x| < 1$. Integrating the series term by term, we obtain

$$
\begin{aligned}
x - \frac{1}{4}x^4 + \frac{1}{7}x^7 - \cdots &= \int_0^x \frac{1}{1+x^3}\,dx \\
&= \frac{1}{3}\ln|1+x| - \frac{1}{6}\ln|x^2 - x + 1| \\
&\quad + \frac{1}{\sqrt{3}}\tan^{-1}\left(\frac{x - 1/2}{\sqrt{3}/2}\right) + \frac{1}{\sqrt{3}}\cdot\frac{\pi}{6}.
\end{aligned}
$$

Evaluating this expression at 1 (see Problem 17.8), we obtain

$$
\sum_{n=1}^{\infty}\frac{(-1)^{n+1}}{3n-2} = \frac{1}{3}\ln 2 + \frac{\sqrt{3}}{9}\pi.
$$

13.4 Solution (1):

$$
\begin{aligned}
\sum_{n=1}^{\infty}\frac{n^3}{n!} &= \sum_{n=1}^{\infty}\frac{n^2}{(n-1)!} \\
&= \sum_{n=0}^{\infty}\frac{(n+1)^2}{n!} \\
&= \sum_{n=0}^{\infty}\frac{n^2}{n!} + 2\sum_{n=0}^{\infty}\frac{n}{n!} + \sum_{n=0}^{\infty}\frac{1}{n!} \\
&= \sum_{n=1}^{\infty}\frac{n}{(n-1)!} + 2\sum_{n=1}^{\infty}\frac{1}{(n-1)!} + e \\
&= \sum_{n=0}^{\infty}\frac{n+1}{n!} + 2\sum_{n=0}^{\infty}\frac{1}{n!} + e \\
&= 2e + 2e + e \\
&= 5e.
\end{aligned}
$$

Solution (2): Since $e^x = \sum_{n=0}^{\infty} \frac{x^n}{n!}$, we have

$$x(x(x(e^x)')')' = \sum_{n=0}^{\infty} \frac{n^3 x^n}{n!}.$$

Hence

$$\sum_{n=0}^{\infty} \frac{n^3}{n!} = e^x + xe^x + 2xe^x + x^2 e^x]_{x=1} = 5e.$$

13.5 Since $(2+\sqrt{3})^{2^{1000}} + (2-\sqrt{3})^{2^{1000}}$ is an integer and $(2-\sqrt{3})^{2^{1000}} < 1$, it follows that

$$[(2+\sqrt{3})^{2^{1000}}] = (2+\sqrt{3})^{2^{1000}} + (2-\sqrt{3})^{2^{1000}} - 1.$$

With this in mind, we define $a_n + b_n\sqrt{3} = (2+\sqrt{3})^{2^n}$, so that $a_0 = 2$, $b_0 = 1$, $a_n = a_{n-1}^2 + 3b_{n-1}^2$, and $b_n = 2a_{n-1}b_{n-1}$. Evidently, $(2+\sqrt{3})^{2^{1000}} + (2-\sqrt{3})^{2^{1000}} = 2a_{1000}$. Therefore, we must calculate a_{1000} modulo 17. From the recurrence relation modulo 17,

$$\{(a_n, b_n)\} = \{(2,1), (7,4), (12,5), (15,1), (7,13),$$
$$(12,12), (15,16), (7,4), \ldots\}.$$

Hence $(a_1, b_1) = (a_7, b_7)$ and $a_{1000} = a_4 = 7$. Therefore the remainder when $[(2+\sqrt{3})]^{2^{1000}}$ is divided by 17 is $2 \cdot 7 - 1 = 13$.

13.6 From the relations

$$x^{2n} - 1 = (x-1)(x+1) \prod_{\substack{k=1 \\ k \neq n}}^{2n-1} \left(x - \cos\frac{k\pi}{n} - i\sin\frac{k\pi}{n}\right)$$

$$= (x-1)(x+1) \cdot \prod_{k=1}^{n-1} \left(x - \cos\frac{k\pi}{n} - i\sin\frac{k\pi}{n}\right)$$

$$\cdot \prod_{k=1}^{n-1} \left(x - \cos\frac{(2n-k)\pi}{n} - i\sin\frac{(2n-k)\pi}{n}\right)$$

$$= (x-1)(x+1) \prod_{k=1}^{n-1} \left(x^2 - 2x\cos\frac{k\pi}{n} + 1\right),$$

it follows that

$$\frac{x^{2n} - 1}{x^2 - 1} = \sum_{k=0}^{n-1} x^{2k} = \prod_{k=1}^{n-1} \left(x^2 - 2x\cos\frac{k\pi}{n} + 1\right).$$

Letting $x = 1$, we obtain

$$n = \prod_{k=1}^{n-1} \left(2 - 2\cos \frac{k\pi}{n} \right)$$

$$= 2^{2n-2} \prod_{k=1}^{n-1} \sin^2 \frac{k\pi}{2n},$$

and therefore

$$\pm\sqrt{n}\, 2^{-n+1} = \prod_{k=1}^{n-1} \sin \frac{k\pi}{2n}.$$

The '+' sign applies because $\sin \frac{k\pi}{2n} > 0$ for all $k = 1, \ldots, n-1$. Similarly,

$$\pm\sqrt{2n+1}\, 2^{-n} = \prod_{k=1}^{n} \sin \frac{k\pi}{2n+1}.$$

For n odd, $n = 2m + 1$,

$$\sin \frac{\pi}{n} \sin \frac{2\pi}{n} \ldots \sin \frac{(n-1)\pi}{n} = \left(\prod_{k=1}^{m} \sin \frac{k\pi}{2m+1} \right)^2$$

$$= \left(\frac{\sqrt{2m+1}}{2^m} \right)^2$$

$$= n 2^{-n+1}.$$

For n even, $n = 2m$,

$$\sin \frac{\pi}{n} \sin \frac{2\pi}{n} \ldots \sin \frac{(n-1)\pi}{n} = \left(\prod_{k=1}^{m-1} \sin \frac{k\pi}{2m} \right)^2$$

$$= \left(\frac{\sqrt{m}}{2^{m-1}} \right)^2$$

$$= n 2^{-n+1}.$$

13.7 First, using integration by parts, if $a \geq b \geq 1$, then

$$\int_0^1 x^a (\ln x)^b \, dx = \frac{-b}{a+1} \int_0^1 x^a (\ln x)^{b-1} \, dx.$$

Hence for $n \geq 1$,

$$\int_0^1 x^n (\ln x)^n \, dx = \frac{-n}{n+1} \int_0^1 x^n (\ln x)^{n-1} \, dx$$

$$= \frac{-n}{n+1} \cdot \frac{-(n-1)}{n+1} \cdot \int_0^1 x^n (\ln x)^{n-2} \, dx$$

$$\vdots$$

$$= \frac{-n}{n+1} \cdot \frac{-(n-1)}{n+1} \cdot \, \cdots \, \cdot \frac{(-1)}{n+1} \cdot \int_0^1 x^n \, dx$$

$$= \frac{(-1)^n n!}{(n+1)^{n+1}}.$$

Therefore,

$$\int_0^1 x^x \, dx = \int_0^1 e^{x \ln x} \, dx$$

$$= \int_0^1 1 + x \ln x + \frac{x^2 (\ln x)^2}{2!} + \cdots \, dx$$

$$= 1 - \frac{1}{2^2} + \frac{1}{3^2} - \cdots$$

$$= \sum_{k=1}^{\infty} \frac{(-1)^{k+1}}{k^k}.$$

We can integrate term by term because the series converges uniformly. This can be shown by the Weierstrass M-test (Glossary). For

$$\frac{(x \ln x)^n}{n!} \leq \frac{M^n}{n!},$$

where M is an upper bound for $x \ln x$ on the interval $(0, 1]$, and the series $\sum_{n=1}^{\infty} \frac{M^n}{n!}$ converges (to $e^M - 1$).

Additional Problems

13.8 Find recursive definitions for the following sequences:

(a) $\{5n - 2\}_{n=1}^{\infty}$;

(b) $\{\frac{n!}{2^n}\}_{n=1}^{\infty}$;

(c) $\{0, 1, 0, 2, 0, 4, 0, 8, \ldots\}$;

(d) $\{1, 1, 4, 10, 28, 76, \ldots\}$.

13.9 Find formulas for the following sequences and use induction to show that the formulas are correct:

(a) $a_n = a_{n-1} + 2n - 1$ for $n \geq 2$, and $a_1 = 4$;

(b) $a_n = 2a_{n-1} - a_{n-2} + 2$ for $n \geq 3$, and $a_1 = 1$ and $a_2 = 4$;

(c) $a_n = a_{n-2} + 4\sqrt{a_{n-2}} + 4$ for $n \geq 2$, and a_0 and $a_1 = 1$.

13.10 Evaluate

$$\sum_{k=0}^{\infty} \frac{1}{(3k)!}$$

in closed form.

13.11 Prove that

$$\sum_{k=1}^{n} k\binom{n}{k} = n2^{n-1}.$$

13.12 (Putnam Competition, 1984) Express

$$\sum_{k=1}^{\infty} \frac{6^k}{(3^{k+1} - 2^{k+1})(3^k - 2^k)}$$

as a rational number.

13.13 Suppose that $a_1 = a_2 = 1$ and $a_n = 3a_{n-1} - a_{n-2}$ for $n > 2$. Prove that none of the numbers $a_1, a_2, \ldots$ is divisible by a prime of the form $4n + 3$.

13.14 Evaluate

$$\sum_{r=1}^{\infty} \left(\sum_{s=1}^{r} s^2\right)^{-1}.$$

13.15 (a) Suppose that b_n is a solution of the recurrence relation

$$a_n = c_1 a_{n-1} + \cdots + c_k a_{n-k} + f(n),$$

for $n \geq k$. (We call b_n a *particular solution* of the recurrence relation.) Let c_n be the general solution to the relation. Show that $d_n = b_n - c_n$ is a solution of the homogeneous recurrence relation

$$a_n = c_1 a_{n-1} + \cdots + c_k a_{n-k}.$$

(b) Verify that $g(n) = n!$ is a solution of the recurrence relation

$$a_n = 8a_{n-1} - 15a_{n-2} + (n-2)! \cdot (n^2 - 9n + 23), \quad n \geq 2.$$

(c) Find the general solution of the homogeneous recurrence relation

$$a_n = 8a_{n-1} - 15a_{n-2}.$$

(d) Find the general solution of the relation in (b).

(e) Find the solution of the relation in (b) for which $a_0 = 2$ and $a_1 = 8$.

13.16 Let $a_n = |a_{n-1}| - a_{n-2}$ for $n \geq 2$, where a_0 and a_1 are given real numbers. Show that this sequence is periodic, with least period exactly 9 unless $a_0 = a_1 = 0$, in which case the least period is 1.

13.17 (Putnam Competition, 1984) Let R be the region consisting of all triples (x, y, z) of nonnegative real numbers satisfying $x + y + z \leq 1$. Let $w = 1 - x - y - z$. Express the volume of the triple integral

$$\iiint_R x^1 y^9 z^8 w^4 \, dx \, dy \, dz$$

in the form $a!b!c!d!/n!$, where a, b, c, d, and n are positive integers.

13.18 Show that

$$\sum_{n=0}^{\infty} f(n)x^n = \frac{1}{(1-x)(1-x^2)(1-x^3)},$$

where $f(n)$ is the integer nearest to $(n+3)^2/12$.

Hint: Use partial fractions and the expansions

$$\frac{1}{1-x} = 1 + x + x^2 + \cdots + x^n + \cdots$$

$$\frac{1}{(1-x)^2} = 1 + 2x + 3x^2 + \cdots + nx^{n-1} + \cdots$$

$$\frac{1}{(1-x)^3} = 1 + 3x + 6x^2 + \cdots + \frac{n(n-1)}{2}x^{n-2} + \cdots$$

$$\frac{1}{1+x+x^2} = \sum_{k=0}^{\infty} \left(x^{3k} - x^{3k+1} \right).$$

13.19 (Putnam Competition, 1981) Let $B(n)$ be the number of ones in the base two expression for the positive integer n. For example, $B(6) = B(110_2) = 2$ and $B(15) = B(1111_2) = 4$. Determine whether or not

$$\exp\left(\sum_{n=1}^{\infty} \frac{B(n)}{n(n+1)} \right)$$

is a rational number. Here $\exp(x)$ denotes e^x.

13.20 Define a sample space in which all $C(2n, n)$ words of length n in the two symbols a, b occur with equal probability. Let E be the event that as the word is read from left to right the number of a's is at least equal to the number of b's. Find $\Pr(E)$.

13.21 Suppose that an object travels a path in the plane beginning at the point $(0,0)$. At each step, the object moves one unit to the right or one unit up. The object stops when it reaches the line $x = n$ or the line $y = n$ (n is a positive integer). Show that the expected length of the object's path is

$$2n - n\binom{2n}{n}2^{1-2n}.$$

13.22 Show that

$$\prod_{r=1}^{\infty}\left(1 + \frac{\binom{2r}{r}}{2(r+1)\left(2^{2r} - \binom{2r}{r}\right)}\right) = 2.$$

13.23 Suppose that the sequence $\{a_n\}$ satisfies a linear recurrence relation of order k with constant coefficients. Let m be a positive integer and p a nonnegative integer. Show that the sequence $\{a_{mn+p}\}$ also satisfies a linear recurrence relation of order k with constant coefficients.

13.24 Prove that the positive integers cannot be partitioned into a finite number (two or more) of arithmetic progressions with all different common differences.

Chapter 14

Maxima and Minima

It was the big race of the day, though heavier in prestige than prize money, an event in which the sponsor, a newspaper, was getting maximum television coverage for minimum outlay.

DICK FRANCIS
Kit Fielding, *Break In*, 1986

Many problems can be solved by finding and exploiting a suitable maximum or minimum.

Example 14.1. (Putnam Competition, 1979; modified) Let A be a set of n points in the plane and B be a set of n points in the plane (all $2n$ points distinct and no three collinear). Prove that the points of A may be paired with the points in B, each pair joined by a line segment, so that no two line segments intersect.

Solution: There are $n!$ possible pairings of the elements of set A with those of set B. Choose a pairing with minimum sum of lengths of line segments. We claim that such a pairing has no intersecting line segments. For suppose that segments $\overline{a_1 b_1}$ and $\overline{a_2 b_2}$, with $a_1, a_2 \in A$ and $b_1, b_2 \in B$, intersect at a point x (see Figure 14.1). Then replacing $\overline{a_1 b_1}$ and $\overline{a_2 b_2}$ by $\overline{a_1 b_2}$ and $\overline{a_2 b_1}$ results in a lesser contribution to the sum of lengths of line segments:

$$|\overline{a_1 b_2}| + |\overline{a_2 b_1}| < |\overline{a_1 x}| + |\overline{x b_2}| + |\overline{a_2 x}| + |\overline{x b_1}| = |\overline{a_1 b_1}| + |\overline{a_2 b_2}|.$$

But this contradicts the assumption that the original pairing had the smallest possible sum of line segment lengths. Therefore the original pairing has no intersecting line segments. ∎

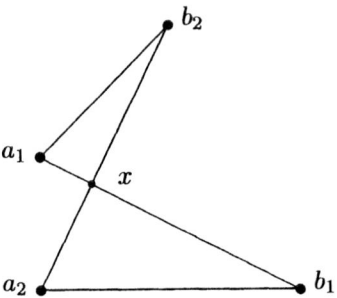

Figure 14.1: Two pairs of line segments.

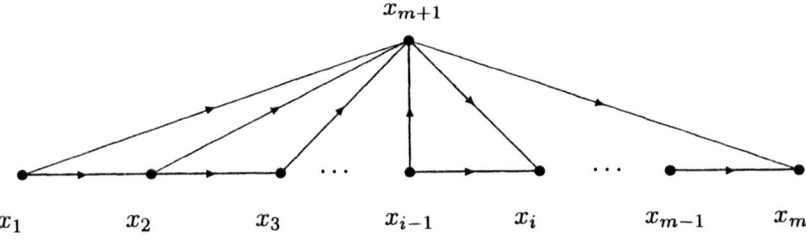

Figure 14.2: Paths in a tournament.

Example 14.2. Show that a tournament (see Glossary) on n vertices contains a directed path of length n.

Solution: We employ both a maximum and a minimum. Suppose that $P = x_1, \ldots, x_m$ is a directed path of greatest length in the tournament and $m < n$. Let x_{m+1} be any vertex not in P. (See Figure 14.2.) By assumption, x_1 is directed to x_{m+1}, and x_{m+1} is directed to x_m. Let i be the least integer such that x_{m+1} is directed to x_i. Then x_{i-1} is directed to x_{m+1}. Therefore $x_1, \ldots, x_{i-1}, x_{m+1}, x_i, \ldots, x_m$ is a directed path of length $m + 1$, contradicting the assumption that P has greatest length. Hence $m = n$, and we are finished. ∎

Problems

14.1 Each of n neighbors has a newsworthy item that they wish to share with each other. How many letters must they send to each other in order to fully inform everyone of all of the news? Suppose that the newsworthy items are distinct from each other, the letters are sent sequentially, and a letter contains all the news known to the writer. Prove that your answer is indeed the minimum.

14.2 Suppose that $f(x) = x^n + a_{n-1}x^{n-1} + \cdots + a_0$ is a polynomial with n real roots. Show that there is at least one root r such that $|r| \geq |a_{n-1}|/n$.

14.3 Find all pairs of positive integers m and n with $m^n = n^m$ and prove that there are no others.

14.4 Show that every finite tournament D contains a vertex v with the following property: for any vertex w of D other than v, there is a directed path of length one or two from v to w.

14.5 (Tarski's fixed point theorem) Let S be a set and $\mathcal{P}(S)$ be the power set of S. Suppose that f maps $\mathcal{P}(S)$ into $\mathcal{P}(S)$ such that whenever $A, B \in \mathcal{P}(S)$ and $A \subseteq B$ then $f(A) \subseteq f(B)$. Show that there exists a set $A \in \mathcal{P}(S)$ with $f(A) = A$.

14.6 (Sylvester's problem) Suppose that A is a finite set of points in the plane with the property that every line determined by two points of A contains a third point of A. Prove that A is a set of collinear points.

14.7 Suppose that f is a continuous function on $[0, 1]$ such that

$$\int_0^1 f(x)\,dx = \int_0^1 xf(x)\,dx = \cdots = \int_0^1 x^{n-1}f(x)\,dx = 0$$

and

$$\int_0^1 x^n f(x)\,dx = 1.$$

Prove that the maximum of $|f(x)|$ on $[0, 1]$ is greater than $2^n(n+1)$.

Solutions

14.1 The minimum is $2n-2$. To see that $2n-2$ communications suffice, let one of the neighbors be designated as Mrs. X. Suppose that each of the other $n - 1$ neighbors in turn sends her news item to Mrs. X and then Mrs. X sends all the accumulated news to each other neighbor in turn. This scheme requires $2n - 2$ communications. (There are other schemes that require only $2n - 2$ letters. For example, number the neighbors 1 through n, and let 1 send her news to 2, who sends all the news she knows to 3, and so on, until n knows all the news.

Then n compiles all the news in one letter and sends copies to all the other neighbors.)

Now we will show that $2n - 2$ is the minimum possible number of letters which fully inform all the neighbors. Since the letters are sent sequentially, there is a first letter which when received gives one neighbor all the news. Let Mrs. Y be the *first* neighbor who possesses all the news. It is evident that at least $n - 1$ communications must occur by the time of this event (one letter is required for each neighbor to tell her original piece of news). Once Mrs. Y has all the news, it takes at least $n - 1$ letters for the others to be fully informed (as each of the other $n - 1$ neighbors must receive a final communication). Therefore at least $2n - 2$ letters are necessary.

14.2 Let the n roots be $r_1, \ldots, r_n$. As

$$(x - r_1) \ldots (x - r_n) = x^n + a_{n-1}x^{n-1} + \cdots + a_0,$$

it follows that $r_1 + \cdots + r_n = -a_{n-1}$. Let r be a root of largest absolute value. Then

$$n|r| \geq |r_1| + \cdots + |r_n| \geq |r_1 + \cdots + r_n| = |a_{n-1}|.$$

The result follows immediately.

14.3 The equation $m^n = n^m$ is obviously satisfied when $m = n$. Assume that $m \neq n$. The given equation is equivalent to $m^{1/m} = n^{1/n}$. Let $f(x) = x^{1/x}$. Note that f is increasing on $[1, e]$ and decreasing on $[e, \infty)$, and also that $f(x) > 1$ when $x > 1$. Therefore one of m or n must equal 2. If $m = 2$, then $n = 4$; and if $n = 2$, then $m = 4$. Hence the solutions for which $m \neq n$ are $(m, n) = (2, 4)$ and $(4, 2)$.

14.4 Let k be a vertex of maximum outdegree d. (The outdegree of a vertex is the number of directed edges leaving that vertex.) Suppose that arrows are directed from k to vertices $x_1, \ldots, x_d$. Suppose that there is some vertex l for which there is not a directed path of length one or two from k to l. Then there are arrows directed from l to vertices $x_1, \ldots, x_d$ and from l to k. Thus the outdegree of l is greater than d, a contradiction.

14.5 Let $\mathcal{C} = \{B \in \mathcal{P}(S) : B \subseteq f(B)\}$ and $A = \bigcup_{B \in \mathcal{C}} B$. We will show that $f(A) = A$ by showing that $A \subseteq f(A)$ and $f(A) \subseteq A$. Let $x \in A$. Then $x \in B$ where $B \in \mathcal{C}$, i.e., $B \subseteq f(B)$. Hence $x \in f(B)$. Now $B \subseteq A$, so $f(B) \subseteq f(A)$ and $x \in f(A)$. Hence $A \subseteq f(A)$. Now $f(A) \subseteq f(f(A))$, so $f(A) \in \mathcal{C}$. Therefore $f(A) \subseteq A$ and we conclude that $f(A) = A$.

14.6 Suppose that A is not a set of collinear points. Let us indicate the line containing points x and y by xy. For each pair of points $\{x, y\} \in A$ and each point z not lying on xy, determine the distance from z to this line. Let x, y, z be such a triple of points with this distance, d, as small as possible. Suppose that the altitude from z to xy is perpendicular to xy at point p. Because the distance from z to xy is minimal, p lies between x and y. By the hypothesis of the problem, xy contains an additional point of A. Call this point w. By the same reasoning as above, w cannot lie between x and y. Without loss of generality, suppose that y lies between x and w. Then the distance from y to zw is less than d, a contradiction.

14.7 Let M be the maximum value of $|f(x)|$ on $[0, 1]$. Then

$$1 = \left| \int_0^1 (x - \frac{1}{2})^n f(x) \, dx \right|$$

$$< M \cdot \int_0^1 \left| x - \frac{1}{2} \right|^n dx$$

$$= M \left[\int_0^{\frac{1}{2}} \left(\frac{1}{2} - x \right)^n dx + \int_{\frac{1}{2}}^1 \left(x - \frac{1}{2} \right)^n dx \right]$$

$$= \frac{M}{(n+1) \cdot 2^n},$$

and hence $M > (n+1) \cdot 2^n$.

Additional Problems

14.8 Which is the larger quantity, e^π or π^e ?

14.9 Let X be a nonempty finite set, and let f be a function mapping X to X. Show that there is at least one nonempty subset S of X such that $f(S) = S$.

14.10 Show that if every plane cross-section of a bounded solid figure is a circle, then the figure is a sphere.

14.11 (Dirac's theorem) A *Hamiltonian circuit* is a path in a graph which visits every vertex exactly once. Let G be a graph with n vertices. Show that if the degree of each vertex of G is greater than $n/2$, then G has a Hamiltonian circuit.

14.12 Evaluate
$$\lim_{n \to \infty} \left[\int_0^{\pi/3} (\sin x + \cos x)^n \, dx \right]^{1/n}.$$

Hint: See the solution to Problem 5.5.

14.13 Show that the vertices of any finite graph can be colored green and red so that at least half of the neighbors of every green vertex are red and at least half of the neighbors of every red vertex are green.

14.14 (*Math. Magazine*, Problem 926, January 1975; modified) A swimmer can swim with speed v in still water. He is required to swim for a given length of time T in a stream whose speed is $r < v$. He is also required to start and finish at the same point. Show that the longest possible path he can complete has length $T(v^2 - r^2)^{1/2}$, and that such a path goes back and forth perpendicular to the stream flow. (Assume that the path is continuous with piecewise continuous first derivative.)

Chapter 15

Means, Inequalities and Convexity

"This affair must all be unraveled from within."

AGATHA CHRISTIE
Hercule Poirot, *The Mysterious Affair at Styles*, 1920

In this chapter we consider problems involving means and inequalities, using the helpful unifying principle of convexity.

There are many proofs of the arithmetic mean–geometric mean (AM–GM) inequality,

$$\frac{a+b}{2} \geq \sqrt{ab}, \qquad (15.1)$$

for real numbers a, $b > 0$. One proof is simply to note that

$$\frac{a+b}{2} = \sqrt{ab} + \left(\frac{\sqrt{a}-\sqrt{b}}{\sqrt{2}}\right)^2.$$

Since the square of any number is nonnegative, the inequality (15.1) follows. But this proof is not easily generalized. A more revealing proof begins by observing that the function $f(x) = -\ln x$ is convex ("lies below its secants") on the interval $(0, \infty)$; see Figure 15.1. Therefore,

$$f\left(\frac{a+b}{2}\right) \leq \frac{f(a)+f(b)}{2},$$

$$-\ln\left(\frac{a+b}{2}\right) \leq \frac{-\ln a - \ln b}{2},$$

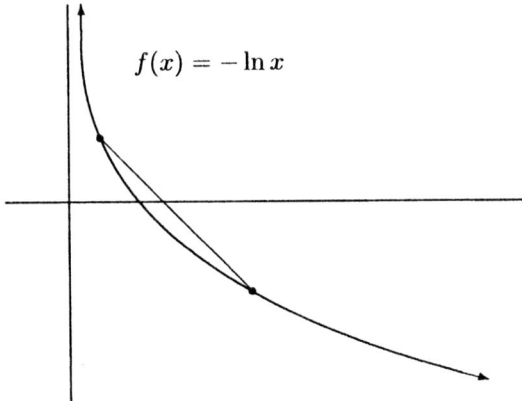

$$f(x) = -\ln x$$

Figure 15.1: A convex function and a secant.

$$\ln\left(\frac{a+b}{2}\right) \geq \frac{\ln a + \ln b}{2},$$

$$\ln\left(\frac{a+b}{2}\right) \geq \ln(ab)^{1/2},$$

and, as the natural logarithm is an increasing function,

$$\frac{a+b}{2} \geq (ab)^{1/2}.$$

This proof generalizes, as we will show. First we define convexity more carefully. A real-valued function f is *convex* on an interval I if

$$f(\lambda a + (1-\lambda)b) \leq \lambda f(a) + (1-\lambda)f(b) \qquad (15.2)$$

for all $a, b \in I$ and $0 \leq \lambda \leq 1$.

The following theorem provides a simple test for convexity of a twice-differentiable function.

Theorem 15.1. If $f''(x) \geq 0$ for all $x \in I$, then f is convex on I.

Proof. Suppose that $a, b \in I$. If $a = b$ there is nothing to prove. So assume that $a < b$, and define

$$d(x) = \frac{f(b) - f(a)}{b - a}(x - a) + f(a) - f(x).$$

Geometrically, $d(x)$ is the vertical distance between $(x, f(x))$ and the secant joining $(a, f(a))$ and $(b, f(b))$. As $d''(x) = -f''(x) \leq 0$, the absolute minimum of d on the interval $[a, b]$ occurs at a or b. Since $d(a) = d(b) = 0$, it follows that $d(x) \geq 0$ for all $x \in [a, b]$. Let $x = \lambda a + (1 - \lambda)b$. Then

$$\frac{f(b) - f(a)}{b - a}[\lambda a + (1 - \lambda)b - a] + f(a) - f(\lambda a + (1 - \lambda)b) \geq 0,$$

and hence

$$\lambda f(a) + (1 - \lambda)f(b) \geq f(\lambda a + (1 - \lambda)b).$$

∎

The next theorem asserts that convex functions satisfy an n-variable version of the inequality (15.2).

Theorem 15.2 (Jensen's inequality). Suppose that f is convex on I. If $a_1, \ldots, a_n \in I$ and $\lambda_1, \ldots, \lambda_n$ are nonnegative real numbers such that $\lambda_1 + \cdots + \lambda_n = 1$, then

$$f\left(\sum_{i=1}^{n} \lambda_i a_i\right) \leq \sum_{i=1}^{n} \lambda_i f(a_i).$$

Proof. The case $n = 2$ holds by definition of convexity. From this fact and the assumption that the relation holds for n,

$$
\begin{aligned}
f\left(\sum_{i=1}^{n+1} \lambda_i a_i\right) &= f\left(\sum_{i=1}^{n} \lambda_i a_i + \lambda_{n+1} a_{n+1}\right) \\
&\leq (1 - \lambda_{n+1})f\left(\frac{1}{(1 - \lambda_{n+1})}\sum_{i=1}^{n} \lambda_i a_i\right) + \lambda_{n+1} f(a_{n+1}) \\
&= (1 - \lambda_{n+1})f\left(\sum_{i=1}^{n} \frac{\lambda_i}{(1 - \lambda_{n+1})} a_i\right) + \lambda_{n+1} f(a_{n+1}) \\
&\leq (1 - \lambda_{n+1})\sum_{i=1}^{n} \frac{\lambda_i}{(1 - \lambda_{n+1})} f(a_i) + \lambda_{n+1} f(a_{n+1}) \\
&= \sum_{i=1}^{n} \lambda_i f(a_i) + \lambda_{n+1} f(a_{n+1}) \\
&= \sum_{i=1}^{n+1} \lambda_i f(a_i).
\end{aligned}
$$

Thus the result holds for $n + 1$ and we are finished. Note that if $\lambda_{n+1} = 0$ above, then the inequality reduces to the n-variable case.   ∎

Example 15.1. Show that

$$\frac{\sin x_1 + \cdots + \sin x_n}{n} \leq \sin\left(\frac{x_1 + \cdots + x_n}{n}\right)$$

for all $x_1, \ldots, x_n \in [0, \pi]$.

Solution: Let $f(x) = -\sin x$. Since $f''(x) \geq 0$ for $x \in [0, \pi]$, f is convex on the interval $[0, \pi]$. The result now follows from Jensen's inequality.   ∎

Example 15.2. Prove the AM–GM inequality for n variables:

$$(a_1 \ldots a_n)^{1/n} \leq \frac{a_1 + \cdots + a_n}{n}.$$

Solution: Let $y = -\ln x$. As $y'' \geq 0$ for $x \in (0, \infty)$, y is convex on the interval $(0, \infty)$. Therefore,

$$\frac{\ln a_1 + \cdots + \ln a_n}{n} \leq \ln\left(\frac{a_1 + \cdots + a_n}{n}\right).$$

Hence

$$\ln\left[(a_1 \ldots a_n)^{1/n}\right] \leq \ln\left(\frac{a_1 + \cdots + a_n}{n}\right).$$

The AM–GM inequality now follows, since ln is an increasing function.

Note: Using the full generality of Theorem 15.2, we obtain a "weighted version" of the AM–GM inequality:

$$(a_1^{w_1} \ldots a_n^{w_n})^{1/w} \leq \frac{w_1 a_1 + \cdots + w_n a_n}{w},$$

where the w_i are nonnegative numbers with sum w.   ∎

We now introduce a generalization of the arithmetic and geometric means. Suppose that $\mathbf{a} = (a_1, \ldots, a_n)$ and $\mathbf{w} = (w_1, \ldots, w_n)$, with $a_i > 0$ and $w_i > 0$ for all $1 \leq i \leq n$. Let $w = \sum_{i=1}^{n} w_i$. Define, for $-\infty \leq r \leq \infty$,

$$M_r = M_r(\mathbf{a}, \mathbf{w}) = \begin{cases} \left(\frac{1}{w}\sum_{i=1}^{n} w_i a_i^r\right)^{1/r} & -\infty < r < \infty, \; r \neq 0 \\[2mm] \left(\prod_{i=1}^{n} a_i^{w_i}\right)^{1/w} & r = 0 \\[2mm] \max_i\{a_i\} & r = \infty \\[2mm] \min_i\{a_i\} & r = -\infty. \end{cases}$$

We call $M_r(\mathbf{a}, \mathbf{w})$ the r-th *power mean* of $\mathbf{a}$ with *weights* $\mathbf{w}$.

The values of M_r with $r = -1$, 0, 1, and 2 are called, respectively, the *harmonic mean* (HM), *geometric mean* (GM), *arithmetic mean* (AM), and *quadratic mean* or *root mean squared* (QM) of **a** with *weights* **w**. That is,

$$HM = \frac{w}{\sum_{i=1}^{n} w_i/a_i}$$

$$GM = \left(\prod_{i=1}^{n} a_i^{w_i}\right)^{1/w}$$

$$AM = \frac{1}{w} \sum_{i=1}^{n} w_i a_i$$

$$QM = \left(\frac{1}{w} \sum_{i=1}^{n} w_i a_i^2\right)^{1/2}.$$

Theorem 15.3. Let **a** and **w** be fixed. Then M_r is a continuous and increasing function of r, for $r \in [-\infty, \infty]$. Moreover, M_r is a strictly increasing function of r unless $a_1 = a_2 = \cdots = a_n$.

Proof. For $0 < r < s < \infty$, let $t = s/r > 1$, and define $f(x) = x^t$ for $x > 0$. As $f''(x) = t(t-1)x^{t-2} \geq 0$, it follows that f is a convex function. Therefore

$$f\left(\frac{1}{w} \sum_{i=1}^{n} w_i a_i^r\right) \leq \frac{1}{w} \sum_{i=1}^{n} w_i f(a_i^r),$$

$$\left(\frac{1}{w} \sum_{i=1}^{n} w_i a_i^r\right)^{s/r} \leq \frac{1}{w} \sum_{i=1}^{n} w_i a_i^s,$$

and

$$\left(\frac{1}{w} \sum_{i=1}^{n} w_i a_i^r\right)^{1/r} \leq \left(\frac{1}{w} \sum_{i=1}^{n} w_i a_i^s\right)^{1/s}.$$

Hence $M_r \leq M_s$.

For $0 < r < \infty$, we apply the AM–GM inequality to $a_1^r, \ldots, a_n^r$ and obtain

$$\left(\frac{1}{w} \sum_{i=1}^{n} w_i a_i^r\right)^{1/r} \geq \left(\prod_{i=1}^{n} a_i^{w_i}\right)^{1/w}.$$

Hence $M_0 \leq M_r$.

The cases $-\infty < r < s \leq 0$ are covered by the above results and the identity

$$M_r((a_1, \ldots, a_n), \mathbf{w}) = \left[M_{-r} \left(\left(\frac{1}{a_1}, \ldots, \frac{1}{a_n} \right), \mathbf{w} \right) \right]^{-1}.$$

If $0 < r < \infty$, then

$$M_r \leq \left(\frac{1}{w} \sum_{i=1}^{n} w_i \max\{a_i\}^r \right)^{1/r} = \max\{a_i\},$$

and hence $M_r \leq M_\infty$. Similarly, $M_{-\infty} \leq M_r$ if $-\infty < r < 0$.

Therefore M_r is an increasing function of r. It is not difficult to show that in all of the above arguments there is strict inequality unless $a_1 = a_2 = \cdots = a_n$.

To demonstrate that M_r is a continuous function for $r \in [-\infty, \infty]$, it suffices to show that $\lim_{r \to \infty} M_r = M_\infty$, $\lim_{r \to -\infty} M_r = M_{-\infty}$, and $\lim_{r \to 0} M_r = M_0$. (As M_r is a composition of continuous functions for $r \in (-\infty, 0)$ and $r \in (0, \infty)$, it is clear that M_r is continuous on these intervals.) To prove the first of these relations, we apply the "squeeze principle" to M_r. We showed above that $M_r \leq M_\infty$. Since

$$M_r \geq \left(\frac{1}{w} \cdot w_i \cdot M_\infty^r \right)^{1/r} = \left(\frac{w_i}{w} \right)^{1/r} \cdot M_\infty$$

and $\lim_{r \to \infty} \left[\left(\frac{w_i}{w} \right)^{1/r} \cdot M_\infty \right] = M_\infty$, it follows that $\lim_{r \to \infty} M_r = M_\infty$. Hence M_r is continuous at ∞. A similar proof shows that M_r is continuous at $-\infty$. Finally, the continuity of M_r at $r = 0$ is proved using l'Hôpital's rule. For $-\infty < r < \infty$, $r \neq 0$,

$$\ln M_r = \frac{1}{r} \ln \left(\frac{1}{w} \sum_{i=1}^{n} w_i a_i^r \right).$$

Therefore,

$$\begin{aligned}
\lim_{r \to 0} \ln M_r &= \lim_{r \to 0} \frac{1}{r} \ln \left(\frac{1}{w} \sum_{i=1}^{n} w_i a_i^r \right) \\
&= \lim_{r \to 0} \frac{w}{\sum_{i=1}^{n} w_i a_i^r} \left(\frac{1}{w} \sum_{i=1}^{n} w_i a_i^r \ln a_i \right) \\
&= \frac{1}{w} \sum_{i=1}^{n} w_i \ln a_i
\end{aligned}$$

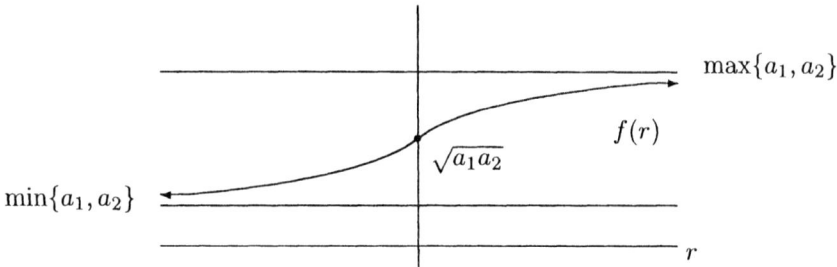

Figure 15.2: The function $f(r) = M_r$ for $n = 2$.

$$= \ln\left(\prod_{i=1}^{n} a_i^{w_i}\right)^{1/w},$$

and

$$\lim_{r \to 0} M_r = \left(\prod_{i=1}^{n} a_i^{w_i}\right)^{1/w} = M_0.$$

Hence M_r is continuous at 0. ∎

Figure 15.2 shows the function $f(r) = M_r$ for two variables a_1, a_2 (and $w_1 = w_2 = \frac{1}{2}$).

Corollary 15.4.

$$\min_i\{a_i\} \le \mathrm{HM} \le \mathrm{GM} \le \mathrm{AM} \le \mathrm{QM} \le \max_i\{a_i\}.$$

Example 15.3. (U. S. A. Olympiad, 1978; modified) Let T be a tetrahedron with three mutually perpendicular edges of lengths a, b, and c. Let l be the sum of the lengths of the six edges of T. What is the maximum possible volume of T?

Solution: Let v be the volume of T. By the AM–GM and QM–GM inequalities,

$$\begin{aligned}
l &= a + b + c + \sqrt{a^2 + b^2} + \sqrt{b^2 + c^2} + \sqrt{a^2 + c^2} \\
&\ge 3(abc)^{1/3} + \sqrt{2ab} + \sqrt{2bc} + \sqrt{2ac} \\
&\ge 3(abc)^{1/3} + 3\sqrt{2}(abc)^{1/3} \\
&= (3 + 3\sqrt{2})(6v)^{1/3},
\end{aligned}$$

and hence

$$v \le \frac{1}{6}\left(\frac{1}{3 + 3\sqrt{2}}\right)^3 l^3,$$

with equality if and only if $a = b = c$.   ∎

We close this chapter by deriving several "classical" inequalities from Theorem 15.3. Specifically, we prove Hölder's inequality, Cauchy's inequality, and Minkowski's inequality.

Let $p > 1$, $q > 1$, and $\frac{1}{p} + \frac{1}{q} = 1$. *Hölder's inequality* is the relation

$$\sum_{i=1}^{n} x_i y_i \leq \left(\sum_{i=1}^{n} x_i^p\right)^{1/p} \left(\sum_{i=1}^{n} y_i^q\right)^{1/q},$$

where $x_i > 0$ and $y_i > 0$ for $1 \leq i \leq n$. To prove it, we start with the inequality

$$\left(\frac{1}{w} \sum_{i=1}^{n} w_i a_i^s\right)^{1/s} \leq \frac{1}{w} \sum_{i=1}^{n} w_i a_i,$$

with $0 < s < 1$ (given by Theorem 15.3), and make the substitutions $s = 1/p$, $a_i = x_i^p/y_i^q$, and $w_i = y_i^q$. This yields

$$\left(\frac{1}{\sum_{i=1}^{n} y_i^q} \sum_{i=1}^{n} x_i y_i\right)^p \leq \frac{1}{\sum_{i=1}^{n} y_i^q} \sum_{i=1}^{n} x_i^p,$$

and Hölder's inequality follows.

Note: In fact, Hölder's inequality is equivalent to the statement that M_r is an increasing function; see Additional Problem 15.22.

The special case $p = q = 2$ of Hölder's inequality is *Cauchy's inequality*:

$$\sum_{i=1}^{n} x_i y_i \leq \left(\sum_{i=1}^{n} x_i^2\right)^{1/2} \left(\sum_{i=1}^{n} y_i^2\right)^{1/2}.$$

Note: Cauchy's inequality can also be proved as follows:

$$\left(\sum_{i=1}^{n} x_i^2\right)\left(\sum_{i=1}^{n} y_i^2\right) - \left(\sum_{i=1}^{n} x_i y_i\right)^2 = \sum_{i,j}(x_i^2 y_j^2 - x_i y_i x_j y_j)$$

$$= \frac{1}{2}\sum_{i,j}(x_i^2 y_j^2 + x_j^2 y_i^2 - 2x_i y_i x_j y_j)$$

$$= \frac{1}{2}\sum_{i,j}(x_i y_j - x_j y_i)^2$$

$$\geq 0.$$

In this proof it is clear that equality occurs only when $x_i y_j = x_j y_i$ for all i, j; that is, when $y_i = k x_i$, $1 \le i \le n$, for some constant k.

From Hölder's inequality we obtain

$$\sum_{i=1}^{n}(x_i + y_i)^r = \sum_{i=1}^{n} x_i (x_i + y_i)^{r-1} + \sum_{i=1}^{n} y_i (x_i + y_i)^{r-1}$$

$$\le \left(\sum_{i=1}^{n} x_i^r\right)^{1/r} \left(\sum_{i=1}^{n}(x_i + y_i)^r\right)^{(r-1)/r}$$

$$+ \left(\sum_{i=1}^{n} y_i^r\right)^{1/r} \left(\sum_{i=1}^{n}(x_i + y_i)^r\right)^{(r-1)/r}$$

$$= \left[\left(\sum_{i=1}^{n} x_i^r\right)^{1/r} + \left(\sum_{i=1}^{n} y_i^r\right)^{1/r}\right]\left(\sum_{i=1}^{n}(x_i + y_i)^r\right)^{(r-1)/r}.$$

Therefore,

$$\left(\sum_{i=1}^{n}(x_i + y_i)^r\right)^{1-(r-1)/r} \le \left(\sum_{i=1}^{n} x_i^r\right)^{1/r} + \left(\sum_{i=1}^{n} y_i^r\right)^{1/r},$$

and

$$\left(\sum_{i=1}^{n}(x_i + y_i)^r\right)^{1/r} \le \left(\sum_{i=1}^{n} x_i^r\right)^{1/r} + \left(\sum_{i=1}^{n} y_i^r\right)^{1/r}.$$

This relation is known as *Minkowski's inequality*.

Question: When does equality occur in Hölder's and Minkowski's relation?

Example 15.4. Given that $\sum_{n=1}^{\infty} a_n$ is convergent and each $a_n \ge 0$, prove that $\sum_{n=1}^{\infty} \sqrt{a_n} \cdot n^{-p}$ converges for $p > 1/2$.

Solution: Consider the partial sums $s_k = \sum_{n=1}^{k} \sqrt{a_n} \cdot n^{-p}$. By Cauchy's inequality,

$$s_k \le \left(\sum_{n=1}^{k} a_n\right)^{1/2} \cdot \left(\sum_{n=1}^{k} n^{-2p}\right)^{1/2}$$

and the result follows since $\sum_{n=1}^{\infty} a_n$ and $\sum_{n=1}^{\infty} n^{-2p}$ are both convergent.

Note: The result is false for $0 < p \le 1/2$. Let

$$a_n = \frac{1}{n(\log n)^2}$$

for $n \geq 2$. Then using the integral test, we see that $\sum_{n=2}^{\infty} a_n$ is convergent but

$$\sum_{n=2}^{\infty} \sqrt{a_n} \cdot n^{-p} \geq \sum_{n=2}^{\infty} \sqrt{a_n} \cdot n^{-1/2} = \sum_{n=2}^{\infty} \frac{1}{n \log n},$$

which diverges. ∎

Problems

15.1 Let $x_1 + x_2 + x_3 = \frac{\pi}{2}$, where the x_i are positive. Show that

$$\sin x_1 \sin x_2 \sin x_3 \leq \frac{1}{8}.$$

15.2 (*Green Book*, Problem 15) Let $a_1, \ldots, a_n$ be given real numbers, which are not all zero. Determine the least value of $x_1^2 + \cdots + x_n^2$, where $x_1, \ldots, x_n$ are real numbers satisfying $a_1 x_1 + \cdots + a_n x_n = 1$.

15.3 Show that $\tan a + \tan b \geq 2 \tan(ab)^{1/2}$ for all $a, b \in [0, \pi/2)$.

15.4 (U. S. A. Olympiad, 1980) Given $0 \leq a, b, c \leq 1$ show that

$$\frac{a}{b+c+1} + \frac{b}{c+a+1} + \frac{c}{a+b+1} + (1-a)(1-b)(1-c) \leq 1.$$

15.5 (U. S. A. Olympiad, 1978; modified) Suppose that $a, b, c, d,$ and e are real numbers such that

$$a + b + c + d + e = 8$$

and

$$a^2 + b^2 + c^2 + d^2 + e^2 = 16.$$

Determine the maximum value of e.

15.6 (Putnam Competition, 1978) Let $p(x) = 2 + 4x + 3x^2 + 5x^3 + 3x^4 + 4x^5 + 2x^6$. For k with $0 < k < 5$, define

$$I_k = \int_0^{\infty} \frac{x^k}{p(x)} \, dx.$$

For which k is I_k smallest?

15.7 Let $p_1, \ldots, p_n$ be distinct points in the closed unit disc in the plane. Let d_k be the distance between p_k and the nearest other such point. Show that $\sum_{k=1}^{n} d_k^2 \leq 16$.

15.8 Given a_0 and b_0 with $0 < a_0 \le b_0$, define a_n and b_n inductively by $a_{n+1} = \sqrt{a_n b_n}$ and $b_{n+1} = (a_n + b_n)/2$, for $n \ge 0$. Prove that the sequences $\{a_n\}$ and $\{b_n\}$ converge to the same limit.

Note: The common limit is known as *Gauss' arithmetic–geometric mean*.

15.9 For n positive real numbers with minimum m and maximum M, let A and G denote their arithmetic and geometric means. Prove that

$$A - G \ge n^{-1} \left(\sqrt{M} - \sqrt{m} \right)^2.$$

15.10 Let $a_1, \ldots, a_n$ be positive real numbers, and let s denote their sum. Show that

$$(1 + a_1)(1 + a_2) \ldots (1 + a_n) \le 1 + \frac{s}{1!} + \frac{s^2}{2!} + \cdots + \frac{s^n}{n!}.$$

15.11 Suppose that P_1, P_2, P_3 are three distinct points on a circle of radius 1. If $d(P_i, P_j)$ denotes the straight line distance from P_i to P_j, prove that the quantity

$$\frac{1}{d(P_1, P_2)} + \frac{1}{d(P_2, P_3)} + \frac{1}{d(P_3, P_1)}$$

is minimized when P_1, P_2, P_3 are the vertices of an equilateral triangle.

15.12 Prove that for any real a, b, c, d, we have

$$\max\{|ac|, |ad + bc|, |bd|\} \ge \left(\frac{\sqrt{5} - 1}{2} \right) \max\{|a|, |b|\} \max\{|c|, |d|\}.$$

Solutions

15.1 By the AM–GM inequality and by convexity,

$$(\sin x_1 \sin x_2 \sin x_3)^{1/3} \le \frac{\sin x_1 + \sin x_2 + \sin x_3}{3} \le \sin \left(\frac{x_1 + x_2 + x_3}{3} \right).$$

Since $x_1 + x_2 + x_3 = \frac{\pi}{2}$,

$$\sin x_1 \sin x_2 \sin x_3 \le \sin^3 \frac{\pi}{6} = \frac{1}{8}.$$

15.2 By Cauchy's inequality,

$$(x_1^2 + \cdots + x_n^2)(a_1^2 + \cdots + a_n^2) \geq (a_1 x_1 + \cdots + a_n x_n)^2 = 1.$$

Hence

$$(x_1^2 + \cdots + x_n^2) \geq (a_1^2 + \cdots + a_n^2)^{-1},$$

with equality if and only if $x_i = a_i(a_1^2 + \cdots + a_n^2)^{-1}$ for each i.

15.3 The function $\tan x$ is increasing and convex for $x \in [0, \frac{\pi}{2})$. Therefore, by the AM–GM inequality,

$$\frac{\tan a + \tan b}{2} \geq \tan\left(\frac{a+b}{2}\right)$$

$$\geq \tan\sqrt{ab}.$$

15.4 Let

$$f(a,b,c) = \frac{a}{b+c+1} + \frac{b}{c+a+1} + \frac{c}{a+b+1} + (1-a)(1-b)(1-c).$$

Holding b and c constant and differentiating with respect to a, we find

$$\frac{\partial^2}{\partial a^2}f = 2b(c+a+1)^{-3} + 2c(a+b+1)^{-3} \geq 0.$$

Therefore f is a convex function of a on the interval $[0,1]$, and f assumes its maximum at one of the endpoints of this interval. The same reasoning shows that $f(a,b,c)$ assumes its maximum when (a,b,c) is a vertex of the cube $[0,1]^3$. Since $f(a,b,c) = 1$ at these vertices, the inequality is demonstrated.

15.5 By Cauchy's inequality,

$$(a \cdot 1 + b \cdot 1 + c \cdot 1 + d \cdot 1)^2 \leq (a^2 + b^2 + c^2 + d^2)(1 + 1 + 1 + 1).$$

Hence

$$(8-e)^2 \leq 4(16 - e^2).$$

It follows that $0 \leq e \leq 16/5$. The upper bound is attained when $a = b = c = d = 6/5$.

15.6 Let $y = 1/x$. We extend the definition of I_k to $-1 < k < 5$, and obtain

$$I_k = \int_0^\infty \frac{x^k}{p(x)}\,dx$$

$$= \int_0^\infty \frac{y^{4-k}}{p(y)}\,dy.$$

$$= I_{4-k}.$$

Hence

$$I_k = \frac{1}{2}(I_k + I_{4-k})$$

$$= \frac{1}{2}\int_0^\infty \frac{x^k + x^{4-k}}{p(x)}\,dx$$

$$\geq \int_0^\infty \frac{x^2}{p(x)}\,dx$$

$$= I_2.$$

15.7 Place a disc of radius $\frac{1}{2}d_k$ around each point p_k. These discs are nonoverlapping and lie inside a disc of radius 2 concentric with the given unit disc. Therefore,

$$\sum_{k=1}^n \pi\left(\frac{1}{2}d_k\right)^2 \leq \pi 2^2,$$

from which the result follows.

Note: The same inequality holds if the points are elements of the square $[-1,1] \times [-1,1]$. Equality occurs with two points at opposite vertices of the square and with four points at the four vertices of the square. See [20, page 27]. For points in the unit disc, the maximum value of $\sum_{k=1}^n d_k^2$ is 9, and this is achieved by three points equally spaced on the disc's circumference.

15.8 By the AM–GM inequality, we have

$$a_{n+1} = \sqrt{a_n b_n} \leq \frac{a_n + b_n}{2} = b_{n+1}$$

for $n \geq 0$, and hence $a_n \leq b_n$ for $n \geq 0$.

Therefore

$$a_{n+1} = \sqrt{a_n b_n} \geq \sqrt{a_n a_n} = a_n$$

and

$$b_{n+1} = \frac{a_n + b_n}{2} \leq \frac{b_n + b_n}{2} = b_n,$$

i.e., $\{a_n\}$ is decreasing and $\{b_n\}$ is increasing. Also, $a_n \leq b_n \leq b_0$, so $\{a_n\}$ is bounded above, Similarly, $\{b_n\}$ is bounded below.

Hence there exist L_1 and L_2 so that $\lim_{n\to\infty} a_n = L_1$ and $\lim_{n\to\infty} b_n = L_2$. From the recurrence relation $b_{n+1} = (a_n + b_n)/2$, it follows that $L_2 = (L_1 + L_2)/2$, which implies that $L_1 = L_2$.

15.9 Let the n positive real numbers be $x_1, \ldots, x_n$ and assume that $x_1 = m$ and $x_n = M$. The inequality $A - G \geq (\sqrt{M} - \sqrt{m})^2/n$ is equivalent to

$$\sum_{j=1}^{n} x_j - n \left(\prod_{j=1}^{n} x_j \right)^{1/n} \geq M - 2\sqrt{Mm} + m$$

or

$$\sum_{j=2}^{n-1} x_j + 2\sqrt{Mm} \geq n \left(\prod_{j=1}^{n} x_j \right)^{1/n}.$$

Applying the AM–GM inequality to the n numbers $\sqrt{Mm}$, x_2, x_3, $\ldots, x_{n-1}$, $\sqrt{Mm}$, we get

$$\sum_{j=2}^{n-1} x_j + 2\sqrt{Mm} \geq n \cdot \left(\prod_{j=2}^{n-1} x_j \cdot \sqrt{Mm}\sqrt{Mm} \right)^{1/n}$$

$$= n \cdot \left(\prod_{j=1}^{n} x_j \right)^{1/n},$$

as needed.

15.10 By the AM–GM inequality,

$$\left[\prod_{j=1}^{n} (1 + a_j) \right]^{1/n} \leq \frac{1}{n} \sum_{j=1}^{n} (1 + a_j) = \frac{1}{n}(n + s) = 1 + \frac{s}{n},$$

so

$$\prod_{j=1}^{n} (1+a_j) \leq \left(1 + \frac{s}{n} \right)^n = \sum_{j=0}^{n} \binom{n}{j} \frac{s^j}{n^j} = \sum_{j=0}^{n} \frac{n!}{(n-j)! \cdot n^j} \cdot \frac{s^j}{j!} \leq \sum_{j=0}^{n} \frac{s^j}{j!}.$$

15.11 Let P_1 and P_2 be fixed points and P_3 a variable point on the circle. Let $\alpha = m(\angle P_1 P_2 P_3)$, $\beta = m(\angle P_2 P_1 P_3)$, and $\gamma = \pi - \alpha - \beta$. Since γ is a constant, by the law of sines (Glossary) we have

$$\frac{\sin \alpha}{d(P_1 P_3)} = \frac{\sin \beta}{d(P_2 P_3)} = k,$$

where k is a constant. Hence

$$f(\alpha, \beta) = \frac{1}{d(P_1 P_3)} + \frac{1}{d(P_2 P_3)} = k \left(\frac{1}{\sin \alpha} + \frac{1}{\sin \beta} \right).$$

Therefore, by the AM–HM inequality and convexity,

$$\frac{2k}{f(\alpha,\beta)} = \frac{2}{\frac{1}{\sin\alpha} + \frac{1}{\sin\beta}} \leq \frac{\sin\alpha + \sin\beta}{2} \leq \sin\left(\frac{\alpha+\beta}{2}\right) = \cos\frac{\gamma}{2}.$$

Equality is achieved if and only if $\alpha = \beta$. It follows that if the angles α, β, and γ are not all equal, then two of the points P_1, P_2, P_3 can be fixed and the third changed to give a lesser value of f. Therefore the minimum value of

$$\frac{1}{d(P_1, P_2)} + \frac{1}{d(P_2, P_3)} + \frac{1}{d(P_3, P_1)}$$

occurs when α, β, and γ are equal.

15.12 If $|a| \leq |b|$ and $|c| \leq |d|$, then the result follows easily. The same is true if $|b| \leq |a|$ and $|d| \leq |c|$.

Now suppose that $|a| \leq |b|$ and $|d| \leq |c|$. We must show that

$$\max\{|ac|, |bd|, |ad + bc|\} \geq \frac{\sqrt{5}-1}{2} \cdot |b| \cdot |c|.$$

If $b = 0$, then $a = 0$, and the inequality reduces to $0 \geq 0$, which is true. Hence we may assume that $b \neq 0$. Similarly, we assume that $c \neq 0$. The inequality to be proved is equivalent to

$$\max\left\{\left|\frac{a}{b}\right|, \left|\frac{d}{c}\right|, \left|\frac{ad}{bc} + 1\right|\right\} \geq \frac{\sqrt{5}-1}{2},$$

or

$$\max\{|x|, |y|, |xy + 1|\} \geq \frac{\sqrt{5}-1}{2},$$

for $|x| \leq 1$, $|y| \leq 1$.

Suppose, without loss of generality, that $|y| \leq |x| \leq 1$ and that

$$|xy + 1| < \frac{\sqrt{5}-1}{2}.$$

Then

$$\frac{1-\sqrt{5}}{2} < xy + 1 < \frac{\sqrt{5}-1}{2},$$

which implies that

$$\frac{-1-\sqrt{5}}{2} < xy < \frac{\sqrt{5}-3}{2}.$$

Hence

$$|xy| > \frac{3 - \sqrt{5}}{2},$$

from which it follows that

$$|x|^2 \geq |xy| > \frac{3 - \sqrt{5}}{2},$$

and

$$|x| > \sqrt{\frac{3 - \sqrt{5}}{2}} = \frac{\sqrt{5} - 1}{2}.$$

The final possibility, $|b| \leq |a|$ and $|c| \leq |d|$, is similar to the above case.

Additional Problems

15.13 Let $x, y \in (0, \infty)$. Show that

$$\log \frac{x^3 + y^3}{2} \geq 3 \log \frac{2xy}{x + y}.$$

Note: Use Theorem 15.3.

15.14 This problem and the next one outline another approach, known as Cauchy's method, to the AM–GM inequality and convex functions.

(a) As noted earlier, it is easy to prove the AM–GM inequality for two positive numbers. Show by induction that if $a_1, a_2, \ldots, a_{2^m}$ are positive numbers, then

$$\frac{1}{2^m} \sum_{k=1}^{2^m} a_k \geq \left(\prod_{k=1}^{2^m} a_k \right)^{1/2^m}.$$

(b) Suppose that $b_1, b_2, \ldots, b_n$ are positive numbers, with $2^{m-1} \leq n < 2^m$, and let $a_1 = b_1, a_2 = b_2, \ldots, a_n = b_n$, and $a_{n+1} = a_{n+2} = \cdots = a_{2^m} = b$, for $b = (b_1 + \cdots + b_n)/n$. Show that

$$\frac{1}{n} \sum_{k=1}^{n} b_k \geq \left(\prod_{k=1}^{n} b_k \right)^{1/n}.$$

15.15 Say that f is *mid-point convex* if

$$f\left(\frac{x+y}{2}\right) \leq \frac{1}{2}f(x) + \frac{1}{2}f(y)$$

for all $x, y \in I$.

(a) Use Cauchy's method to prove that if f is mid-point convex, then

$$f\left(\frac{x_1 + x_2 + \cdots + x_n}{n}\right) \leq \frac{f(x_1) + f(x_2) + \cdots + f(x_n)}{n}$$

for $x_1, x_2, \ldots, x_n \in I$.

(b) Show that if f is mid-point convex and r is any rational number between 0 and 1, then

$$f(rx + (1-r)y) \leq rf(x) + (1-r)f(y)$$

for all $x, y \in I$.

(c) Show that if f is mid-point convex and continuous on I, then f is convex on I.

15.16 We say that a real-valued function f is *subadditive* if $f(x+y) \leq f(x) + f(y)$ for all $x, y \in \mathbf{R}$ and *positive homogeneous* if $f(tx) = tf(x)$ for all $t \geq 0$ and all $x \in \mathbf{R}$. Prove that every subadditive, positive homogeneous function is convex.

15.17 Let n be a positive integer. For $k = 0, 1, 2, \ldots, 2n - 2$ define

$$I_k = \int_0^\infty \frac{x^k}{x^{2n} + x^n + 1} \, dx.$$

Prove that $I_k \geq I_{n-1}$, for all k. Show that $I_{n-1} = 2\sqrt{3}\pi/9n$.

Hint: As in the solution to Problem 15.6, make the change of variables $y = 1/x$.

15.18 (Putnam Competition, 1978) Let $0 < x_i < \pi$ for $i = 1, 2, \ldots, n$ and set

$$x = \frac{x_1 + x_2 + \cdots + x_n}{n}.$$

Prove that

$$\prod_{i=1}^{n} \frac{\sin x_i}{x_i} \leq \left(\frac{\sin x}{x}\right)^n.$$

15.19 Suppose that g is an increasing convex function on $\mathbf{R}$ and f is a convex function of $\mathbf{R}$. Prove that $g \circ f$ is a convex function, where $g \circ f(x) = g(f(x))$.

15.20 Let A_n and G_n be the arithmetic and geometric means, respectively, of
$$\binom{n}{0}, \ldots, \binom{n}{n}.$$
Prove that $\lim_{n\to\infty} A_n^{1/n} = 2$ and $\lim_{n\to\infty} G_n^{1/n} = e^{1/2}$.

15.21 Let $f(r) = r \ln M_r$. Prove that f is a convex function of r.

Hint: Let $\lambda_1 > 0$ and $\lambda_2 > 0$ with $\lambda_1 + \lambda_2 = 1$. Use Hölder's inequality with exponents $1/\lambda_1$ and $1/\lambda_2$.

15.22 (a) Prove Hölder's inequality using convexity.

Hint: Let $p > 1$, $q > 1$, and $\frac{1}{p} + \frac{1}{q} = 1$. Then $f(x) = x^q$ is a convex function, so
$$\left(\sum_{i=1}^{n} \lambda_i x_i\right)^q \le \sum_{i=1}^{n} \lambda_i x_i^q.$$
Now let $\lambda_i = a_i^p / \sum_{i=1}^{n} a_i^p$ and $x_i = b_i a_i^{1-p}$, and obtain
$$\left(\sum_{i=1}^{n} \frac{a_i^p}{\sum_{i=1}^{n} a_i^p} b_i a_i^{1-p}\right)^q \le \sum_{i=1}^{n} \frac{a_i^p}{\sum_{i=1}^{n} a_i^p} \left(b_i a_i^{1-p}\right)^q.$$

Hölder's inequality may be deduced from this relation.

(b) Using Hölder's inequality, prove the statement that M_r is an increasing function of r.

Hint: Assume that $r < s$. In Hölder's inequality, let $p = s/r$, $q = s/(s-r)$, $x_i = a_i^r$ and $y_i = 1$ for all i.

15.23 (M. Klamkin) Suppose that $(b_1, \ldots, b_n)$ is a permutation of $(a_1, \ldots, a_n)$, where $a_i > 0$ for $i = 1, \ldots, n$, and $r, s \ge 0$. Show that
$$\frac{a_1^{r+s}}{b_1^s} + \cdots + \frac{a_n^{r+s}}{b_n^s} \ge a_1^r + \cdots + a_n^r.$$

15.24 (*Math. Magazine*, Problem 1322, June 1989) An n-gon of consecutive sides a_1, a_2, $\ldots$, a_n is circumscribed about a circle of unit radius. Determine the minimum value of the product of all its sides.

Chapter 16

Mean Value Theorems

"You can ... never foretell what any one man will do, but you can say with precision what an average number will be up to."

SIR ARTHUR CONAN DOYLE
Sherlock Holmes, *The Sign of Four*, 1890

In this chapter, we consider problems which have solutions involving "mean value theorems." The four most basic mean value theorems are actually all easily derived from each other and are usually referred to as Rolle's theorem, the mean value theorem (MVT), Cauchy's mean value theorem, and Taylor's theorem.

Theorem 16.1 (Rolle's theorem). If a real-valued function f is continuous on $[a, b]$ and differentiable in (a, b), and if $f(a) = f(b)$, then $f'(c) = 0$ for at least one c in (a, b).

Proof. We first note that the result is trivial if f is constant on $[a, b]$. If f is not constant then there exists an x in (a, b) for which $f(x) \neq f(a) = f(b)$. If $f(x) > f(a)$ then the absolute maximum for f on $[a, b]$ is assumed at some point c in the open interval (a, b), and is therefore also a relative maximum, and so necessarily $f'(c) = 0$. The case $f(x) < f(a)$ is similar. ■

By applying Theorem 16.1 to the function

$$g(x) = f(x) - \left[f(a) + \frac{f(b) - f(a)}{b - a}(x - a) \right],$$

we obtain the following theorem.

Theorem 16.2 (Mean value theorem). If f is continuous on $[a, b]$ and differentiable in (a, b), then there exists c in (a, b) for which

$$f'(c) = \frac{f(b) - f(a)}{b - a}.$$

We note that in the mean value theorem, if $f(a) = f(b)$, the result coincides with that of Rolle's theorem.

If Theorem 16.1 is applied to the function

$$h(x) = [f(b) - f(a)]g(x) - [g(b) - g(a)]f(x),$$

the result is the following theorem.

Theorem 16.3 (Cauchy's mean value theorem). If f and g are continuous on $[a, b]$ and differentiable in (a, b), then there exists c in (a, b) for which

$$[f(b) - f(a)]g'(c) = [g(b) - g(a)]f'(c).$$

Note that in Cauchy's theorem, if either of the functions f or g is the identity function $i(x) = x$, then the conclusion is the same as that of the mean value theorem.

Theorems 16.1 through 16.3 all may be interpreted geometrically as saying that for any smooth continuous curve in the plane, the tangent line at some point along the curve is parallel to the line joining the endpoints of the curve.

Theorem 16.4 (Taylor's theorem). Suppose that $f^{(n+1)}$ exists on $[a, b]$ and $a \leq c \leq b$. Then for each x in $[a, b]$, there exists a point z between c and x for which

$$f(x) = f(c) + f'(c)(x - c) + \frac{f''(c)(x - c)^2}{2!} + \cdots$$

$$+ \frac{f^{(n)}(c)(x - c)^n}{n!} + \frac{f^{(n+1)}(z)(x - c)^{n+1}}{(n + 1)!}.$$

A proof of Theorem 16.4 can be obtained by applying Theorem 16.3 to the functions

$$F(t) = f(x) - \sum_{k=0}^{n} \frac{f^{(k)}(t)(x - t)^k}{k!} \quad \text{and} \quad G(t) = (x - t)^{n+1},$$

defined on the interval $c \leq t \leq x$ (or $x \leq t \leq c$ as the case may be).

Also, we note that in the case $n = 0$, $c = a$, $x = b$, Taylor's theorem reduces to the mean value theorem.

Example 16.1. Evaluate

$$\lim_{n\to\infty} \left(n \cot \frac{\pi}{4n}\right)^{-\frac{1}{2}} \sum_{j=1}^{n} \sqrt{1 + \tan \frac{j\pi}{4n} \tan \frac{(j-1)\pi}{4n}}.$$

Solution: The numbers $x_j = j\pi/4n$, $0 \leq j \leq n$, give the regular partition of the interval $[0, \pi/4]$ into subintervals of length $\pi/4n$. Also, using the identity

$$\tan(A - B) = \frac{\tan A - \tan B}{1 + \tan A \tan B},$$

we have

$$\tan \frac{\pi}{4n} \left[1 + \tan \frac{j\pi}{4n} \tan \frac{(j-1)\pi}{4n} \right] = \tan \frac{j\pi}{4n} - \tan \frac{(j-1)\pi}{4n},$$

which by Theorem 16.2 is equal to $\sec^2(c_j) \cdot \pi/4n$ for some $c_j \in (x_{j-1}, x_j)$. Hence

$$\lim_{n\to\infty} \left(n \cot \frac{\pi}{4n}\right)^{-\frac{1}{2}} \sum_{j=1}^{n} \sqrt{1 + \tan \frac{j\pi}{4n} \tan \frac{(j-1)\pi}{4n}}$$

$$= \lim_{n\to\infty} n^{-\frac{1}{2}} \sum_{j=1}^{n} \sqrt{\tan \frac{\pi}{4n} \left[1 + \tan \frac{j\pi}{4n} \tan \frac{(j-1)\pi}{4n} \right]}$$

$$= \lim_{n\to\infty} n^{-\frac{1}{2}} \sum_{j=1}^{n} \sqrt{\tan \frac{j\pi}{4n} - \tan \frac{(j-1)\pi}{4n}}$$

$$= \lim_{n\to\infty} n^{-\frac{1}{2}} \sum_{j=1}^{n} \sec c_j \cdot \frac{\sqrt{\pi}}{2\sqrt{n}}$$

$$= \frac{2}{\sqrt{\pi}} \lim_{n\to\infty} \sum_{j=1}^{n} \sec c_j \cdot \frac{\pi}{4n}$$

$$= \frac{2}{\sqrt{\pi}} \int_{0}^{\frac{\pi}{4}} \sec x \, dx$$

$$= \frac{2}{\sqrt{\pi}} \ln(\sqrt{2} + 1).$$

■

Example 16.2. Show that $e^x > 3x^2/2$ for $x \geq 0$.

Solution: For $x > 0$, Taylor's theorem gives

$$e^x = 1 + x + \frac{x^2}{2} + \frac{x^3}{6} + e^c \cdot \frac{x^4}{24},$$

where $0 < c < x$. Therefore,

$$e^x > 1 + x + \frac{x^2}{2} + \frac{x^3}{6} + \frac{x^4}{24}$$

and hence $e^x/x^2 > f(x)$ where

$$f(x) = \frac{1}{x^2} + \frac{1}{x} + \frac{1}{2} + \frac{x}{6} + \frac{x^2}{24}.$$

That $f(x) > 3/2$ is straightforward to check by considering which of the intervals

$(0, \frac{3}{2})$, $[\frac{3}{2}, 2)$, $[2, \frac{5}{2})$, $[\frac{5}{2}, 3)$, $[3, \frac{7}{2})$, or $[\frac{7}{2}, \infty)$ contains x.   ■

The next group of mean value theorems involve integrals.

Theorem 16.5 (First mean value theorem for integrals). If f is continuous on $[a, b]$ then there exists a point c between a and b for which

$$\int_a^b f(x)\, dx = f(c) \cdot (b - a).$$

Proof. Let m and M denote respectively the absolute minimum and maximum values of f on $[a, b]$. Integrating the inequality $m \leq f(x) \leq M$ over the interval $[a, b]$ yields

$$m \leq \frac{1}{b - a} \int_a^b f(x)\, dx \leq M,$$

and so by the intermediate value theorem,

$$\frac{1}{b - a} \int_a^b f(x)\, dx = f(c)$$

for some c between a and b.   ■

We now use Cauchy's mean value theorem to derive the following theorem.

Theorem 16.6. If f and g are continuous on $[a,b]$, then there exists a point c in (a,b) for which

$$f(c)\int_a^b g(t)\,dt = g(c)\int_a^b f(t)\,dt.$$

Proof. Let $G(x) = \int_a^x g(t)\,dt$ and $F(x) = \int_a^x f(t)\,dt$. Then Theorem 16.3 implies that there exists c in (a,b) for which

$$F'(c)[G(b) - G(a)] = G'(c)[F(b) - F(a)],$$

or equivalently

$$f(c)\int_a^b g(t)\,dt = g(c)\int_a^b f(t)\,dt \ .$$

∎

The following theorem will be used as a lemma for the second mean value theorem.

Theorem 16.7. If f and g are continuous on $[a,b]$ and if g is nonzero on the interval, then there exists c in (a,b) for which

$$\int_a^b f(x)g(x)\,dx = f(c)\int_a^b g(x)\,dx \ .$$

Proof. Applying Theorem 16.6 to the pair of functions $f(x)g(x)$ and $g(x)$, there exists c in (a,b) for which

$$g(c)\int_a^b f(x)g(x)\,dx = f(c)g(c)\int_a^b g(x)\,dx \ .$$

∎

Theorem 16.8 (Second mean value theorem for integrals). Suppose that g and f' are continuous on $[a,b]$ with f' never zero on the interval. Then there exists c in (a,b) for which

$$\int_a^b f(x)g(x)\,dx = f(a)\int_a^c g(x)\,dx + f(b)\int_c^b g(x)\,dx \ .$$

Proof. Let $G(x) = \int_a^x g(t)\,dt$. Integration by parts gives

$$\int_a^b f(x)g(x)\,dx = f(b)G(b) - \int_a^b f'(x)G(x)\,dx.$$

By Theorem 16.7,

$$\int_a^b f'(x)G(x)\,dx = G(c)\int_a^b f'(x)\,dx = G(c)[f(b) - f(a)]$$

for some c in (a, b).

Hence

$$
\begin{aligned}
\int_a^b f(x)g(x)\,dx &= f(b)G(b) - G(c)[f(b) - f(a)] \\
&= f(a)G(c) + f(b)[G(b) - G(c)] \\
&= f(a)\int_a^c g(x)\,dx + f(b)\int_c^b g(x)\,dx.
\end{aligned}
$$

∎

Example 16.3. Let $0 < a < b$. Show that

$$\left| \int_a^b \sin(x^2)\,dx \right| \le \frac{1}{a} + \frac{1}{b}.$$

Solution: Expressing the integral as $\int_a^b \frac{1}{2x} 2x \sin(x^2)\,dx$ and applying Theorem 16.8, there exists c in (a, b) for which

$$
\begin{aligned}
\int_a^b \sin(x^2)\,dx &= \frac{1}{2a}\int_a^c 2x\sin(x^2)\,dx + \frac{1}{2b}\int_c^b 2x\sin(x^2)\,dx \\
&= \frac{1}{2a}[\cos(a^2) - \cos(c^2)] + \frac{1}{2b}[\cos(c^2) - \cos(b^2)],
\end{aligned}
$$

and hence

$$
\begin{aligned}
\left| \int_a^b \sin(x^2)\,dx \right| &\le \frac{1}{2a}|\cos(a^2) - \cos(c^2)| + \frac{1}{2b}|\cos(c^2) - \cos(b^2)| \\
&\le \frac{1}{2a}\cdot 2 + \frac{1}{2b}\cdot 2 \\
&= \frac{1}{a} + \frac{1}{b}
\end{aligned}
$$

∎

Example 16.4. Let $r < 1$. Show that

$$\lim_{x\to\infty} x^r \int_x^{x+1} \sin(t^2)\,dt = 0.$$

Solution: As in Example 16.3,

$$\left|\int_x^{x+1} \sin(t^2)\,dt\right| \le \frac{1}{x} + \frac{1}{x+1} = \frac{2x+1}{x^2+x},$$

and therefore

$$\left|x^r \int_x^{x+1} \sin(t^2)\,dt\right| \le \frac{2x^{r+1}+x^r}{x^2+x} \to 0$$

as $x \to \infty$. ∎

Problems

16.1 Given that f is increasing on $[0,1]$, $f(0) = 0$, f' exists and is increasing in $(0,1)$, prove that $g(x) = f(x)/x$ is increasing in $(0,1)$.

16.2 Let f be defined on $[0,1]$ with continuous second derivative, and suppose that $f(0) = f(1) = 0$, and $|f''(x)| \le 1$ for $0 \le x \le 1$. Show that $\int_0^1 f(x)\,dx \le 1/12$ and that this is best possible.

16.3 Suppose that a, b, c are real constants with $a > 0$, $c > 0$, and $b^2 \le 3ac$. Show that the equation $\cos x = ax^3 + bx^2 + cx$ has a unique solution in the interval $(0, \pi/2)$.

16.4 Let $f(x)$ be continuous on the closed interval $[a, b]$ and differentiable in the open interval (a, b). If $f(a) = f(b) = 0$, show that for any nonzero real number r there is a number $c \in (a, b)$ such that $r\cdot f'(c)+f(c) = 0$.

16.5 (Putnam Competition, 1981) Find

$$\lim_{t\to\infty}\left[e^{-t}\int_0^t\int_0^t \frac{e^x - e^y}{x-y}\,dx\,dy\right]$$

or show that the limit does not exist.

16.6 Let $z_1, z_2, \ldots, z_n$ be arbitrary complex numbers. Show that there is a subset J of $\{1,\ldots,n\}$ such that $|\sum_{j\in J} z_j| \ge \frac{1}{\pi}\sum_{j=1}^n |z_j|$. Show that $1/\pi$ is the largest constant that works for all n.

16.7 Prove the following assertions:

(a) If f is continuous and strictly increasing on $[a, b]$, then

$$\int_a^b f(x)\, dx + \int_{f(a)}^{f(b)} f^{-1}(x)\, dx = bf(b) - af(a).$$

(b) If f is continuous and strictly increasing on $[0, \infty)$ with $f(0) = 0$ and $\lim_{x \to \infty} f(x) = \infty$, then for any two positive numbers a and b,

$$ab \le \int_0^a f(x)\, dx + \int_0^b f^{-1}(x)\, dx.$$

(c) If a, b, p, q are positive numbers with $\frac{1}{p} + \frac{1}{q} = 1$, then

$$ab \le \frac{a^p}{p} + \frac{b^q}{q}.$$

Solutions

16.1 For $0 < x < 1$,

$$g'(x) = \frac{xf'(x) - f(x)}{x^2}.$$

By the mean value theorem,

$$f(x) = f(x) - f(0) = f'(c)(x - 0) = xf'(c)$$

for some c in $(0, x)$. Since f' is increasing, $f'(x) - f'(c) > 0$, and hence

$$g'(x) = \frac{f'(x) - f'(c)}{x} > 0.$$

Since $g' > 0$ in $(0, 1)$, g is increasing in $(0, 1)$.

16.2 Note that for $g(x) = \frac{x}{2} - \frac{x^2}{2}$, we have $g(0) = g(1) = 0$, $|g''(x)| = 1$ for $0 \le x \le 1$, and

$$\int_0^1 g(x)\, dx = \left(\frac{x^2}{4} - \frac{x^3}{6} \right) \Bigg]_0^1 = \frac{1}{12}.$$

Therefore the solution will be complete if we show that $f(x) \le g(x)$ for $0 \le x \le 1$. Define

$$h(x) = g(x) - f(x).$$

Then $h(0) = h(1) = 0$ and $h''(x) = -1 - f''(x) \leq 0$ for $0 \leq x \leq 1$ (since $|f''(x)| \leq 1$). It follows that h is nonnegative on $[0, 1]$. To see this clearly, suppose that $h(a) < 0$ for some a, $0 < a < 1$. Then, by the mean value theorem,

$$h(a) = h(a) - h(0) = h'(x_1) \cdot a$$

for some x_1, $0 < x_1 < a$, and also,

$$-h(a) = h(1) - h(a) = h'(x_2) \cdot (1 - a)$$

for some x_2, $a < x_2 < 1$. Hence $x_1 < x_2$, $h'(x_1) = h(a)/a < 0$, and $h'(x_2) = h(a)/(a - 1) > 0$, so $h'(x_1) < h'(x_2)$. This contradicts the fact that h' is nonincreasing (since $h'' \leq 0$).

Therefore, $h(x) \geq 0$ or equivalently, $f(x) \leq \frac{x}{2} - \frac{x^2}{2}$ for $0 \leq x \leq 1$.

16.3 Let $f(x) = ax^3 + bx^2 + cx - \cos x$. Then $f(0) = -1$ and

$$
\begin{aligned}
f\left(\frac{\pi}{2}\right) &= \frac{a\pi}{2}\left[\frac{\pi^2}{4} + \frac{b\pi}{2a} + \frac{c}{a}\right] \\
&= \frac{a\pi}{2}\left[\left(\frac{\pi}{2} + \frac{b}{2a}\right)^2 + \frac{4ac - b^2}{4a^2}\right] \\
&> \frac{a\pi}{2}\left[\left(\frac{\pi}{2} + \frac{b}{2a}\right)^2 + \frac{3ac - b^2}{4a^2}\right] \\
&\geq 0.
\end{aligned}
$$

Therefore, by the intermediate value theorem, the equation has at least one solution in $(0, \pi/2)$.

Now if there were more than one such solution then Rolle's theorem would imply a solution of $f'(x) = 0$ in $(0, \pi/2)$. But

$$
\begin{aligned}
f'(x) &= 3ax^2 + 2bx + c + \sin x \\
&> 3ax^2 + 2bx + c \\
&= 3a\left[x^2 + \frac{2b}{3a}x + \frac{c}{3a}\right] \\
&= 3a\left[\left(x + \frac{b}{3a}\right)^2 + \frac{3ac - b^2}{9a^2}\right] \\
&\geq 0.
\end{aligned}
$$

16.4 Define $g(x) = e^{x/r} \cdot f(x)$. Then g is continuous on $[a, b]$, $g(a) = g(b) = 0$, and g is differentiable in (a, b) with

$$
\begin{aligned}
g'(x) &= e^{x/r} \cdot f'(x) + (1/r) \cdot e^{x/r} \cdot f(x) \\
&= (1/r) \cdot e^{x/r} \cdot [r \cdot f'(x) + f(x)].
\end{aligned}
$$

By Rolle's theorem, there exists c in (a, b) such that $g'(c) = 0$, that is, such that $r \cdot f'(c) + f(c) = 0$.

16.5 Let

$$
f(t) = \int_0^t \int_0^t \frac{e^x - e^y}{x - y} \, dx \, dy.
$$

Note that by Theorem 16.2, for $x \neq y$ we have

$$
\frac{e^x - e^y}{x - y} = e^z
$$

for some z between x and y, and hence

$$
f(t) \geq \int_0^t \int_0^t dx \, dy = t^2,
$$

so $\lim_{t \to \infty} f(t) = \infty$. Thus l'Hôpital's rule is appropriate to find $\lim_{t \to \infty} \frac{f(t)}{e^t}$. Now

$$
\begin{aligned}
\frac{f(t+h) - f(t)}{h} &= \frac{\int_0^{t+h} \int_0^{t+h} \frac{e^x - e^y}{x-y} \, dx \, dy - \int_0^t \int_0^t \frac{e^x - e^y}{x-y} \, dx \, dy}{h} \\
&= \frac{2 \int_0^t \int_t^{t+h} \frac{e^x - e^y}{x-y} \, dx \, dy + \int_t^{t+h} \int_t^{t+h} \frac{e^x - e^y}{x-y} \, dx \, dy}{h}.
\end{aligned}
$$

By Theorem 16.5,

$$
\frac{1}{h} \int_t^{t+h} \frac{e^x - e^y}{x - y} \, dx = \frac{e^z - e^y}{z - y}
$$

for some z between t and $t + h$. Consequently,

$$
f'(t) = \lim_{h \to 0} \frac{f(t+h) - f(t)}{h} = 2 \int_0^t \frac{e^t - e^y}{t - y} \, dy.
$$

Therefore,

$$
\frac{f'(t)}{e^t} = 2 \int_0^t \frac{1 - e^{y-t}}{t - y} \, dy.
$$

Making the change of variables $u = t - y$, this gives

$$\frac{f'(t)}{e^t} = 2 \int_0^t \frac{1 - e^{-u}}{u}\, du > 2 \int_1^t \frac{1 - e^{-u}}{u}\, du > 2(1 - e^{-1})\ln t.$$

Thus

$$\lim_{t \to \infty} \frac{f(t)}{e^t} = \lim_{t \to \infty} \frac{f'(t)}{e^t} = \infty.$$

16.6 For each θ, $0 \le \theta \le 2\pi$, let $S_\theta = \{j : \operatorname{Im} z_j e^{-i\theta} \ge 0\}$ and $f(\theta) = \sum_{j \in S_\theta} \operatorname{Im} z_j e^{-i\theta}$. This amounts to rotating the complex number coordinate system by the angle θ and then taking only the parts of the z_i that lie above the new x-axis. Now $f(\theta) \le |\sum_{j \in S_\theta} z_j|$. Furthermore,

$$\int_0^{2\pi} f(\theta)\, d\theta = \sum_{j=1}^n \int_{\operatorname{Im} z_j e^{-i\theta} \ge 0} \operatorname{Im} z_j e^{-i\theta}\, d\theta = \sum_{j=1}^n 2|z_j|.$$

(Recall that $\operatorname{Im} z = |z| \sin \arg z$.) Hence

$$\frac{1}{2\pi} \int_0^{2\pi} f(\theta)\, d\theta = \frac{1}{\pi} \sum_{j=1}^n |z_j|.$$

By the mean value theorem for integrals, there exists θ such that $f(\theta) = \frac{1}{\pi} \sum_{j=1}^n |z_j|$. Therefore θ determines the appropriate complex numbers.

Now we show that the constant $1/\pi$ cannot be increased. Let $z_1, \ldots, z_n$ be the sides of a regular n-gon, oriented clockwise. As $n \to \infty$, the quantity $\sum_{j=1}^n |z_j|$ approaches $2\pi r$, where r is the radius of the n-gon's circumscribed circle. Clearly, a subset of the vectors can have length no greater than $2r$, the diameter of the circle. Therefore, for this choice of vectors,

$$\lim_{n \to \infty} \frac{|\sum_{j \in J} z_j|}{\sum_{j=1}^n |z_j|} \le \frac{1}{\pi}.$$

16.7 (a) Letting $x = f(t)$ in the second integral we have

$$\int_{f(a)}^{f(b)} f^{-1}(x)\, dx = \int_a^b t\, df(t).$$

Then using the integration by parts formula for Riemann-Stieltjes integrals,

$$\int_a^b t\, df(t) = bf(b) - af(a) - \int_a^b f(t)\, dt.$$

(b) Case 1: $f(a) \leq b$. Using (a),

$$\int_0^a f(x)\,dx + \int_0^b f^{-1}(x)\,dx = \int_0^a f(x)\,dx + \int_0^{f(a)} f^{-1}(x)\,dx$$
$$+ \int_{f(a)}^b f^{-1}(x)\,dx$$
$$= af(a) + \int_{f(a)}^b f^{-1}(x)\,dx.$$

Now by Theorem 16.5,

$$\int_{f(a)}^b f^{-1}(x)\,dx = f^{-1}(c)[b - f(a)]$$

for some c where $f(a) \leq c \leq b$, and hence $f^{-1}(c) \geq a$. Thus we have

$$\int_0^a f(x)\,dx + \int_0^b f^{-1}(x)\,dx = af(a) + f^{-1}(c)[b - f(a)]$$
$$\geq af(a) + a[b - f(a)]$$
$$= ab.$$

Case 2: If $b \leq f(a)$, then

$$\int_0^a f(x)\,dx + \int_0^b f^{-1}(x)\,dx = af(a) - \int_b^{f(a)} f^{-1}(x)\,dx$$
$$= af(a) - f^{-1}(c)[f(a) - b]$$

where $b \leq c \leq f(a)$. So $f^{-1}(c) \leq a$ and hence again

$$\int_0^a f(x)\,dx + \int_0^b f^{-1}(x)\,dx \geq af(a) - a[f(a) - b] = ab.$$

(c) Noting that $\frac{1}{p} + \frac{1}{q} = 1$ implies that $\frac{1}{p-1} = q - 1$, it follows that if $f(x) = x^{p-1}$, then $f^{-1}(x) = x^{q-1}$. So, by (b), we have

$$\frac{a^p}{p} + \frac{b^q}{q} = \int_0^a f(x)\,dx + \int_0^b f^{-1}(x)\,dx \geq ab.$$

Additional Problems

16.8 (Putnam Competition, 1950; modified) Suppose that x_0 is a real number. For $k \geq 1$, let $x_k = \cos x_{k-1}$. Show that $\lim_{k \to \infty} x_k$ exists and is independent of x_0.

16.9 Suppose that f and f' are continuous on the interval (a, b) and $|f'(x)| \leq M$ for x in (a, b). If c is in (a, b) and $f(c) > 0$, show that $f(x) > 0$ for any x in (a, b) such that $|x - c| < f(c)/M$.

16.10 Show that the power series representation of the function

$$f(x) = \sum_{n=0}^{\infty} \frac{x^n (1 - x)^{2n}}{n!}$$

cannot have three consecutive zero coefficients. (Actually it has none.)

16.11 (*Math. Magazine*, Problem 1303, October 1988) Find all continuous functions f on $(0, \infty)$ such that

$$\int_x^{x^2} f(t)\ dt = \int_1^x f(t)\ dt$$

for all $x > 0$.

16.12 (*American Math. Monthly*, Problem E3214, June–July 1987) Let f be a real function with $n + 1$ derivatives on $[a, b]$. Suppose that $f^{(i)}(a) = f^{(i)}(b) = 0$ for $i = 0, 1, \ldots, n$. Prove that there is a number ξ in (a, b) such that $f^{(n+1)}(\xi) = f(\xi)$.

16.13 (*Math. Magazine*, Problem 1053, September 1978) Let $f(x)$ be differentiable on $[0, 1]$ with $f(0) = 0$ and $f(1) = 1$. For each positive integer n, show that there exist distinct $x_1, x_2, \ldots, x_n$ such that

$$\sum_{i=1}^{n} \frac{1}{f'(x_i)} = n.$$

16.14 (*Math. Magazine*, Problem 1060, January 1979) Prove or disprove: There exists a function f defined on $[-1, 1]$ with f'' continuous such that $\sum_{n=1}^{\infty} f(1/n)$ converges but $\sum_{n=1}^{\infty} |f(1/n)|$ diverges.

16.15 (*Math. Magazine*, Problem 950, September 1975) Show that there is a unique real number c such that for every differentiable function f on $[0, 1]$ with $f(0) = 0$ and $f(1) = 1$, the equation $f'(x)(x) = cx$ has a solution in $(0, 1)$.

Chapter 17

Summation by Parts

Shifting problems is the first rule for a long and pleasant life.

AMANDA CROSS
Professor Kate Fansler,
The Theban Mysteries, 1971

Given that the two series $\sum_{n=1}^{\infty} a_n$ and $\sum_{n=1}^{\infty} b_n$ converge, the series $\sum_{n=1}^{\infty} a_n b_n$ may or may not converge. For example, $\sum_{n=1}^{\infty} \frac{(-1)^n}{\sqrt{n}}$ converges but

$$\sum_{n=1}^{\infty} \frac{(-1)^n}{\sqrt{n}} \cdot \frac{(-1)^n}{\sqrt{n}} = \sum_{n=1}^{\infty} \frac{1}{n},$$

which, being the harmonic series, diverges.

However, if $\sum_{n=1}^{\infty} a_n$ converges and $\sum_{n=1}^{\infty} b_n$ is absolutely convergent (Glossary), then $\sum_{n=1}^{\infty} a_n b_n$ is also absolutely convergent. The same result holds if $\{a_n\}$ is merely bounded.

Theorem 17.1. If $\{a_n\}$ is bounded and $\sum_{n=1}^{\infty} b_n$ is absolutely convergent, then $\sum_{n=1}^{\infty} a_n b_n$ is absolutely convergent.

Proof. We are given that $|a_n| \leq M$ for some M. Thus $|a_n b_n| \leq M \cdot |b_n|$, and therefore $\sum_{n=1}^{\infty} a_n b_n$ converges absolutely by the comparison test. ∎

For any sequence $\{a_n\}$, let A_k be its kth partial sum; that is, $A_k = \sum_{i=1}^{k} a_i$.

Theorem 17.2 (Abel's formula). Let $\{a_n\}$ and $\{b_n\}$ be sequences. Then, for $n \geq 2$,

$$\sum_{k=1}^{n} a_k b_k = A_n b_n - \sum_{k=1}^{n-1} A_k (b_{k+1} - b_k).$$

We gave an induction proof of Abel's formula in Chapter 4. We now give a slightly more direct proof.

Proof. We have

$$\sum_{k=1}^{n} a_k b_k \;=\; a_1 b_1 + a_2 b_2 + \cdots + a_n b_n$$

$$=\; A_1 b_1 + (A_2 - A_1) b_2 + \cdots + (A_n - A_{n-1}) b_n$$

$$=\; A_1(b_1 - b_2) + A_2(b_2 - b_3) + \cdots + A_{n-1}(b_{n-1} - b_n) + A_n b_n$$

$$=\; A_n b_n - \sum_{k=1}^{n-1} A_k (b_{k+1} - b_k).$$

∎

The following proposition is a direct consequence of Abel's formula.

Corollary 17.3. If $\{A_n b_n\}$ converges and $\sum_{n=1}^{\infty} A_n(b_{n+1} - b_n)$ converges, then $\sum_{n=1}^{\infty} a_n b_n$ converges.

Theorem 17.4 (Abel's test). If $\sum_{n=1}^{\infty} a_n$ converges and $\{b_n\}$ is monotonic and convergent, then $\sum_{n=1}^{\infty} a_n b_n$ converges.

Proof. Since $\{A_n\}$ and $\{b_n\}$ converge, $\{A_n b_n\}$ converges.

If $\{b_n\}$ is an increasing sequence, then

$$\sum_{n=1}^{\infty} |b_{n+1} - b_n| = \sum_{n=1}^{\infty} (b_{n+1} - b_n),$$

while if $\{b_n\}$ is a decreasing sequence, then

$$\sum_{n=1}^{\infty} |b_{n+1} - b_n| = \sum_{n=1}^{\infty} (b_n - b_{n+1}).$$

In both cases $\sum_{n=1}^{\infty} |b_{n+1} - b_n|$ is convergent. Therefore $\sum_{n=1}^{\infty} (b_{n+1} - b_n)$ is absolutely convergent. Since $\{A_n\}$ is bounded, $\sum_{n=1}^{\infty} A_n(b_{n+1} - b_n)$ is absolutely convergent by Theorem 17.1. Therefore, by Corollary 17.3, $\sum_{n=1}^{\infty} a_n b_n$ converges. ∎

Example 17.1. Suppose that $\sum_{n=1}^{\infty} a_n$ converges. Show that

$$\sum_{n=1}^{\infty} \frac{a_n}{n}$$

converges.

Solution: This follows directly from Abel's test, since $\{\frac{1}{n}\}$ is monotonic. ∎

Theorem 17.5 (Dirichlet's test). If the partial sums of $\sum_{n=1}^{\infty} a_n$ are bounded and $\{b_n\}$ decreases monotonically to zero, then $\sum_{n=1}^{\infty} a_n b_n$ converges.

Proof. Since $b_n \to 0$ and $\{A_n\}$ is bounded, it follows that $A_n b_n \to 0$. Also,

$$\sum_{n=1}^{\infty} |b_{n+1} - b_n| = \sum_{n=1}^{\infty} (b_n - b_{n+1}) = b_1 < \infty.$$

Therefore $\sum_{n=1}^{\infty}(b_{n+1} - b_n)$ is absolutely convergent. By Theorem 17.1, $\sum_{n=1}^{\infty} A_n(b_{n+1} - b_n)$ is also absolutely convergent. By Corollary 17.3, $\sum_{n=1}^{\infty} a_n b_n$ converges. ∎

The alternating series test is a special case of Dirichlet's test, where $a_n = (-1)^{n+1}$.

Theorem 17.6 (Alternating series test). If $\{b_n\}$ decreases to 0, then $\sum_{n=1}^{\infty}(-1)^{n+1}b_n$ converges.

Example 17.2. Use Dirichlet's test to show that if $\sum_{n=1}^{\infty} a_n$ converges, then $\sum_{n=1}^{\infty} a_n/n$ converges.

Solution: Just take $b_n = 1/n$ in Dirichlet's test.

Note: The same result was obtained in Example 17.1. ∎

Example 17.3. Show that

$$\sum_{n=1}^{\infty} \frac{\sin n}{n} < \infty.$$

Solution: We need the formulas

$$\sin z = \frac{e^{iz} - e^{-iz}}{2i} \tag{17.1}$$

and

$$\cos z = \frac{e^{iz} + e^{-iz}}{2}. \tag{17.2}$$

Now suppose that $x \in \mathbf{R}$, $x \neq 2k\pi$. Then

$$\sum_{j=1}^{n} e^{ijx} = e^{ix}\left(\frac{e^{ixn} - 1}{e^{ix} - 1}\right) = \frac{\sin(nx/2)}{\sin(x/2)} e^{(n+1)ix/2}, \tag{17.3}$$

and we obtain

$$\sum_{j=1}^{n}[\cos jx + i\sin jx] = \frac{\sin(nx/2)}{\sin(x/2)}\{\cos[(n+1)x/2] + i\sin[(n+1)x/2]\}.$$

Two formulas follow:

$$\sum_{j=1}^{n}\cos jx = \begin{cases} \frac{\sin(nx/2)}{\sin(x/2)}\cos[(n+1)x/2] & x \neq 2k\pi \\ n & x = 2k\pi \end{cases} \tag{17.4}$$

$$\sum_{j=1}^{n}\sin jx = \begin{cases} \frac{\sin(nx/2)}{\sin(x/2)}\sin[(n+1)x/2] & x \neq 2k\pi \\ n & x = 2k\pi. \end{cases} \tag{17.5}$$

Letting $x = 1$ in the formula (17.5) we see that the partial sums of $\sum_{n=1}^{\infty}\sin n$ are bounded. By Dirichlet's test,

$$\sum_{n=1}^{\infty}\frac{\sin n}{n} < \infty.$$

■

Example 17.4. Determine the set of z values for which

$$\sum_{n=1}^{\infty}\frac{z^n}{n}$$

converges.

Solution: The series converges absolutely for $|z| < 1$ and diverges for $|z| > 1$. Using the equation (17.3) with $z = e^{it}$, $t \neq 2k\pi$, we find by Dirichlet's test that the series converges for all $|z| = 1$ except when $z = 1$. When $z = 1$ the series is the harmonic series and therefore diverges.   ■

Problems

17.1 Use Abel's fomula (Theorem 17.2) to obtain the formulas

   (a)

$$\sum_{i=1}^{n} i = \frac{n(n+1)}{2}$$

   (b)

$$\sum_{i=1}^{n} i^2 = \frac{n(n+1)(2n+1)}{6}$$

(c)

$$\sum_{i=1}^{n} i^3 = \frac{n^2(n+1)^2}{4}.$$

17.2 Show that if $\{a_n\}$ is monotonic and $\sum_{n=1}^{\infty} a_n$ converges, then

$$\sum_{n=1}^{\infty} n(a_n - a_{n+1})$$

converges, and that the two sums are equal.

17.3 Show that

$$\sum_{n=1}^{\infty} \frac{|\sin n|}{n}$$

diverges.

17.4 Suppose that $\sum_{n=1}^{\infty} a_n$ is convergent and suppose that $v_1(x)$, $v_2(x)$, ... are nonnegative valued functions on some set S such that

$$K \geq v_1(x) \geq v_2(x) \geq \cdots$$

for each $x \in S$. Show that $\sum_{n=1}^{\infty} v_n(x)$ converges uniformly for $x \in S$.

17.5 For a positive integer n and a (positive) divisor d of n, let L_d denote the sum of all divisors of n which are less than or equal to d, i.e.,

$$L_d = \sum_{e|n, e \leq d} e.$$

Similarly, let

$$F_d = \sum_{e|n, e \leq d} \frac{1}{e},$$

$$S_d = \sum_{e|n, e \leq d} e^2$$

and

$$t_d = \sum_{e|n, e \leq d} 1.$$

Also, if $1 \leq d < n$, let d^+ be the smallest divisor of n larger than d.

Show that each of the following is a necessary and sufficient condition for n to be a perfect number (see Glossary):

(a)
$$\tau(n) = 2 + \sum_{d<n} L_d \left(\frac{1}{d} - \frac{1}{d^+} \right)$$

(where $\tau(n)$ is the number of divisors of n)

(b)
$$\tau(n) = 2n - \sum_{d<n} F_d(d^+ - d)$$

(c)
$$\sum_{d|n} d^2 = 2n^2 - n \sum_{d<n} S_d \left(\frac{1}{d} - \frac{1}{d^+} \right)$$

(d)
$$2n^2 = 2n + \sum_{d<n} F_d[(d^+)^2 - d^2]$$

(e)
$$\sum_{d|n} d^3 = 2n^3 - \sum_{d<n} L_d[(d^+)^2 - d^2]$$

(f)
$$n[\tau(n) - 2] = \sum_{d<n} t_d(d^+ - d)$$

(g)
$$\sum_{d|n} d^2 = 2n^2 - \sum_{d<n} L_d(d^+ - d).$$

Solutions

17.1 (a) Let $A = \sum_{i=1}^{n} i$. By Abel's formula,

$$A = \sum_{i=1}^{n} 1 \cdot i = n^2 - \sum_{i=1}^{n-1} i[(i+1) - i] = n^2 - \sum_{i=1}^{n-1} i = n^2 - A + n;$$

hence $2A = n^2 + n$ and $A = n(n+1)/2$.

(b) Let $B = \sum_{i=1}^{n} i^2 = \sum_{i=1}^{n} 1 \cdot i^2$. By Abel's formula, $B = n^3 - \sum_{i=1}^{n-1} i[(i+1)^2 - i^2] = n^3 - \sum_{i=1}^{n-1} (2i^2 + i) = n^3 - 2(B - n^2) - n(n-1)/2$. Solving for B gives the result.

The formula in (c) follows similarly.

17.2 Consider only $a_1 \geq a_2 \geq a_3 \geq \cdots \geq a_n \geq \cdots \geq 0$, since $\lim_{n \to \infty} a_n = 0$ ($\sum_{n=1}^{\infty} a_n$ converges).

Let $s_k = \sum_{n=1}^{k} n(a_n - a_{n+1})$. Then

$$
\begin{aligned}
s_k &= a_1 + a_2(2-1) + \cdots + a_k(k-(k-1)) - ka_{k+1} \\
&= \sum_{n=1}^{k} a_n - ka_{k+1}.
\end{aligned}
$$

Therefore, since $\sum_{n=1}^{\infty} a_n$ converges, it follows that

$$
\lim_{n \to \infty} (a_n + a_{n+1} + \cdots + a_{2n}) = 0.
$$

But $a_n + \cdots + a_{2n} \geq na_{2n}$. Thus

$$
\lim_{n \to \infty} na_{2n} = 0,
$$

and hence $\lim_{n \to \infty} na_{n+1} = 0$.

Therefore

$$
\lim_{k \to \infty} s_k = \lim_{k \to \infty} \sum_{i=1}^{k} a_i - \lim_{k \to \infty} (ka_{k+1}) = \lim_{k \to \infty} \sum_{i=1}^{k} a_i,
$$

which converges.

17.3 Solution (1): Note that

$$
\begin{aligned}
\frac{|\sin n|}{n} &\geq \frac{\sin^2 n}{n} \\
&= \frac{1 - \cos 2n}{2n}.
\end{aligned}
$$

Since $\sum_{n=1}^{\infty} \frac{1}{2n}$ diverges and $\sum_{n=1}^{\infty} \frac{\cos 2n}{2n}$ converges (the proof of this is similar to the solution in Example 17.3), it follows that

$$
\sum_{n=1}^{\infty} \frac{|\sin n|}{n}
$$

diverges.

Solution (2): Suppose that $|\sin n| \leq \sin \frac{1}{2}$. Then $|\sin(n+1)| \geq \sin \frac{1}{2}$. Hence at least one of $|\sin n|$ and $|\sin(n+1)|$ is greater than or equal to $\sin \frac{1}{2}$. Therefore

$$
\sum_{n=1}^{\infty} \frac{|\sin n|}{n} \geq \sin \frac{1}{2} \sum_{n=2k}^{\infty} \frac{1}{n} = \frac{\sin \frac{1}{2}}{2} \sum_{n=1}^{\infty} \frac{1}{n}.
$$

Hence the given series diverges.

17.4 Let $f_N(x) = \sum_{n=1}^{N} a_n v_n(x)$. Then, for $N \geq M$,

$$
\begin{aligned}
|f_N(x) - f_M(x)| &= |a_{M+1} v_{M+1}(x) + \cdots + a_N v_N(x)| \\
&\leq K|a_{M+1} + \cdots + a_N| \\
&\leq K\epsilon
\end{aligned}
$$

(using the Cauchy criterion and the fact that $\sum a_n$ converges). Thus $\lim_{n \to \infty} f_N(x)$ converges uniformly.

17.5 By definition, n is perfect if and only if $\sigma(n) = 2n$ where $\sigma(n) = \sum_{d|n} d$. To prove (a), we use Abel's formula (Theorem 17.2), taking a_k to be the kth largest divisor of n and $b_k = 1/a_k$. Thus

$$
\tau(n) = \sum_{d|n} 1 = \sum_{d|n} d \cdot \frac{1}{d} = \sigma(n) \cdot \frac{1}{n} - \sum_{d<n} L_d \left(\frac{1}{d^+} - \frac{1}{d} \right).
$$

The result is then clear since n is perfect if and only if $\sigma(n) \cdot \frac{1}{n} = 2$.

Part (b) follows similarly, only taking b_k to be the kth divisor of n and $a_k = 1/b_k$. Thus

$$
\tau(n) = \sum_{d|n} \frac{1}{d} \cdot d = \left(\sum_{d|n} \frac{1}{d} \right) \cdot n - \sum_{d<n} F_d(d^+ - d)
$$

$$
= \sigma(n) - \sum_{d<n} F_d(d^+ - d).
$$

The other parts are done in similar fashion. For (c), start with $\sigma(n) = \sum_{d|n} d^2 \cdot \frac{1}{d}$ and apply Abel's formula. For (d), start with $\sigma(n) = \sum_{d|n} \frac{1}{d} \cdot d^2$. For (e), $\sum_{d|n} d^3 = \sum_{d|n} d \cdot d^2$; for (f), $\sigma(n) = \sum_{d|n} 1 \cdot d$; for (g), $\sum_{d|n} d^2 = \sum_{d|n} d \cdot d$.

Additional Problems

17.6 Prove that if $\sum_{n=1}^{\infty} a_n$ converges, then each of the following also converge:

(a)

$$
\sum_{n=2}^{\infty} \frac{a_n}{\ln n}
$$

(b)

$$
\sum_{n=1}^{\infty} n^{1/n} a_n
$$

(c)

$$\sum_{n=1}^{\infty} \frac{n+1}{n} a_n$$

(d)

$$\sum_{n=1}^{\infty} \left(1 + \frac{1}{n}\right)^n a_n.$$

17.7 Prove the following two statements:

(a) If $\sum_{n=0}^{\infty} a_n$ is convergent, then $\sum_{n=0}^{\infty} a_n x^n$ converges for $0 \le x \le 1$.

(b) If $\sum_{n=0}^{\infty} a_n$ has bounded partial sums, then $\sum_{n=0}^{\infty} a_n x^n$ converges for $0 \le x < 1$.

17.8 (Abel's theorem) Show that if $\sum_{k=1}^{\infty} a_k$ converges, then

$$\lim_{r \to 1^-} \sum_{k=1}^{\infty} a_k r^k = \sum_{k=1}^{\infty} a_k.$$

Show, however, that the existence of $\lim_{r \to 1^-} \sum_{k=1}^{\infty} a_k r^k$ does not imply the convergence of $\sum_{k=1}^{\infty} a_k$.

Hint: For the second part of the problem, let $a_k = (-1)^k$.

17.9 Prove that for a sequence $\{a_n\}$ and a fixed number $x_0 \in \mathbf{R}$, if $\sum_{n=1}^{\infty} a_n \cdot n^{-x_0}$ converges, then $\sum_{n=1}^{\infty} a_n \cdot n^{-x}$ converges for all $x \ge x_0$.

Hint: Write

$$\sum_{n=1}^{\infty} \frac{a_n}{n^x} = \sum_{n=1}^{\infty} \frac{a_n}{n^{x_0}} \cdot \frac{1}{n^{x-x_0}}.$$

17.10 Suppose that $\sum_{n=0}^{\infty} a_n$ is convergent and that $\{f_n\}_{n=0}^{\infty}$ is a sequence of functions defined on $[a, b]$ such that $0 \le f_{n+1}(x) \le f_n(x) \le M$ holds (for some constant M) for all $n \ge 0$ and $a \le x \le b$. Show that $\sum_{n=0}^{\infty} a_n f_n(x)$ converges uniformly on $[a, b]$.

17.11 (*American Math. Monthly*, Problem E3088, May 1985) Show that for every positive integer n,

$$\sum_{k=1}^{n} \frac{k \cdot k!}{n^k} \binom{n}{k} = n.$$

17.12 (*Math. Magazine*, Problem 1123, May 1981) For which positive integers p is the following result true? If $\{a_n\}$ is a sequence of real numbers and $\sum a_n^p$ converges, then $\sum a_n/n$ must converge.

Chapter 18

Estimation

I was getting closer to the truth. I could feel it in my bones. I felt like I was circling, the orbits getting tighter as I approached the central point.

SUE GRAFTON
Kinsey Millhone, *'E' is for Evidence*, 1988

In this chapter we solve problems by making estimates, often trapping an elusive unknown quantity between upper and lower bounds.

We start by defining some useful notation—symbols called "big oh" and "little oh." Assume that f and g are nonnegative-valued functions defined on the set of positive integers. We say that f is "big oh" of g, and write $f(n) = O(g(n))$, if

$$f(n) \le cg(n)$$

for some positive constant c and all $n \ge n_0$ where n_0 is some nonnegative integer. We say that f is "little oh" of g, and write $f(n) = o(g(n))$, if

$$\lim_{n \to \infty} \frac{f(n)}{g(n)} = 0.$$

One more concept is very helpful. We say that f is "asymptotic" to g, and write $f(n) \sim g(n)$, if

$$\lim_{n \to \infty} \frac{f(n)}{g(n)} = 1.$$

Example 18.1. Estimate

$$\int_{-1}^{1} (1 - x^2)^n \, dx$$

as $n \to \infty$.

Solution: Method (1): Let

$$I_n = \int_{-1}^{1} (1 - x^2)^n \, dx.$$

Then

$$
\begin{aligned}
I_n &= \int_{-1}^{1} e^{n \ln(1-x^2)} \, dx \\
&= \int_{-1}^{1} e^{-nx^2 - nx^4/2 - nx^6/3 - nx^8/4 - \cdots} \, dx.
\end{aligned}
$$

Making the change of variables $x = t/\sqrt{n}$, we obtain

$$
\begin{aligned}
I_n &= \frac{1}{\sqrt{n}} \int_{-\sqrt{n}}^{\sqrt{n}} e^{-t^2 - t^4/2n - t^6/3n^2 - \cdots} \, dt \\
&= \frac{1}{\sqrt{n}} \int_{-\sqrt{n}}^{\sqrt{n}} e^{-t^2} \cdot e^{o(1)} \, dt
\end{aligned}
$$

By Lebesgue's dominated convergence theorem,

$$\lim_{n \to \infty} \sqrt{n} I_n = \int_{-\infty}^{\infty} e^{-t^2} \, dt,$$

which equals $\sqrt{\pi}$ (by Example 19.2). Thus

$$I_n \sim \sqrt{\frac{\pi}{n}}$$

as $n \to \infty$.

Method (2): Using integration by parts, letting $u = (1 - x^2)^n$ and $dv = dx$, we obtain

$$I_n = x(1 - x^2)^n \big|_{-1}^{1} + 2n \int_{-1}^{1} x^2 (1 - x^2)^{n-1} \, dx = 2n I_{n-1} - 2n I_n$$

so that

$$I_n = \frac{2n}{2n + 1} I_{n-1}.$$

It follows by Wallis' formula (Glossary) that

$$I_n \sim \sqrt{\frac{\pi}{n + \frac{1}{2}}} \sim \sqrt{\frac{\pi}{n}}.$$

Method (3): Let

$$I_\lambda = \int_{-1}^{1} (1 - x^2)^\lambda \, dx.$$

We will show that $I_\lambda \sim \sqrt{\pi/\lambda}$ as $\lambda \to \infty$. Since $I_{\lambda-1/2} > I_\lambda > I_{\lambda+1/2}$, it is sufficient to prove that $I_{n-1/2} \sim \sqrt{\pi/n}$ for integral n:

$$\begin{aligned}
I_{n-1/2} &= \int_{-1}^{1} (1 - x^2)^n \frac{dx}{\sqrt{1 - x^2}} \\
&= \frac{1}{2} \int_0^{2\pi} \cos^{2n}(\theta) \, d\theta \\
&= \frac{1}{2} \oint_{S^1} \left(\frac{z + z^{-1}}{2} \right)^{2n} \frac{dz}{iz} \\
&= \frac{1}{2} \frac{2\pi i}{i} \frac{(2n)!}{2^n n! n!}.
\end{aligned}$$

The last step follows by Cauchy's formula (Glossary). The final expression is asymptotic to $\sqrt{\pi/n}$ (by Wallis' formula or Stirling's approximation).

Note: By considering the binomial expansion of $(1 - x^2)^n$, we see that the asymptotic result in this problem yields an estimate of the sum

$$\sum_{k=0}^{n} \frac{(-1)^k}{2k + 1} \binom{n}{k}.$$

Some problems require the use of both induction and analytic techniques (e.g., the AM–GM inequality).

Example 18.2. Find constants $A > 1$, $B > 1$ such that

$$A^{n^2} \le \prod_{k=0}^{n} \binom{n}{k} \le B^{n^2}$$

for all $n \ge 2$.

Solution: We will show

$$2^{n^2/4} \le \prod_{k=0}^{n} \binom{n}{k} \le 2^{n^2}$$

$(A = 2^{1/4}, B = 2)$.

First we show by induction that

$$\frac{n^n}{n!} \geq 2^{(2n-1)/4}$$

for $n \geq 2$. For $n = 2$ we have $2 \geq 2^{3/4}$. If the inequality holds for some $n \geq 2$, then

$$\frac{(n+1)^{n+1}}{(n+1)!} = \frac{(n+1)^n}{n!} \geq \frac{n^n + n \cdot n^{n-1}}{n!} =$$

$$\frac{2n^n}{n!} \geq 2 \cdot 2^{(2n-1)/4} = 2^{(2n+3)/4} > 2^{(2n+1)/4}.$$

Now we use induction to show that

$$2^{n^2/4} \leq \prod_{k=0}^{n} \binom{n}{k}.$$

For $n = 2$, equality holds. If

$$2^{n^2/4} \leq \prod_{k=0}^{n} \binom{n}{k}$$

for some $n \geq 2$, then

$$\prod_{k=0}^{n+1} \binom{n+1}{k} = \prod_{k=0}^{n} \binom{n+1}{k} = \prod_{k=0}^{n} \left[\frac{n+1}{n+1-k} \binom{n}{k} \right] =$$

$$\frac{(n+1)^{n+1}}{(n+1)!} \prod_{k=0}^{n} \binom{n}{k} \geq 2^{(2n+1)/4} \cdot 2^{n^2/4} = 2^{(n+1)^2/4}.$$

To show that

$$\prod_{k=0}^{n} \binom{n}{k} \leq 2^{n^2},$$

we use the AM–GM inequality:

$$\left[\prod_{k=0}^{n} \binom{n}{k} \right]^{1/(n+1)} \leq \frac{\sum_{k=0}^{n} \binom{n}{k}}{n+1} = \frac{2^n}{n+1},$$

and thus

$$\prod_{k=0}^{n} \binom{n}{k} \leq \frac{2^{n(n+1)}}{(n+1)^{n+1}} < \frac{2^{n(n+1)}}{2^n} = 2^{n^2}.$$

■

Example 18.3. (Schütte's theorem) Show that for every positive integer m, there exists a tournament T (see Glossary) such that for each subset $S \subseteq T$, $S = |m|$, there exists a vertex $p \in T - S$ which is directed to each vertex of S.

Solution: Let V be a set of n vertices, where n is to be determined later. We will show that there exists a tournament T of the required kind on V if n is sufficiently large. There are $2^{\binom{n}{2}}$ possible tournaments on V. Given a subset $S \subseteq V$ of m vertices, let A_S be the set of tournaments on n vertices for which there is no vertex $p \in V - S$ directed to each vertex of S. Then

$$\left| \bigcup_{S \subseteq V} A_S \right| \leq \binom{n}{m} |A_S|$$

$$= \binom{n}{m} 2^{n-m}$$

$$< 2^{\binom{n}{2}}.$$

(when n is sufficiently large). The second inequality holds because $\binom{n}{m}$ is a polynomial in n and the exponential function $2^{\binom{n}{2}}$ grows faster than the exponential function 2^{n-m}. Hence there is a tournament of the required kind if n is sufficiently large. ∎

Problems

18.1 Estimate how fast $\sum_{n=0}^{\infty} x^{n^2}$ tends to ∞ as $x \to 1^-$.

18.2 For a sequence $\{x_n\}$, define the sequence $\{y_n\}$ by

$$y_n = \frac{x_1 + \cdots + x_n}{n}.$$

Prove that if $\lim_{n \to \infty} x_n = x$, then $\lim_{n \to \infty} y_n = x$.

18.3 Let $a_1, a_2, \ldots$ be real numbers such that $\sum_{k=1}^{\infty} a_k/k$ converges. Show that

$$\lim_{n \to \infty} \frac{1}{n} \sum_{k=1}^{n} a_k = 0.$$

Solutions

18.1 We use the integral bounds $\sum_{n=0}^{\infty} x^{n^2} \geq \int_0^{\infty} x^{t^2}\, dt$ and $\sum_{n=1}^{\infty} x^{n^2} \leq \int_0^{\infty} x^{t^2}\, dt$. Since

$$\int_0^{\infty} x^{t^2}\, dt = \frac{1}{2}\sqrt{\frac{\pi}{-\ln x}}$$

(using Example 19.2), it follows that

$$\frac{1}{2}\sqrt{\frac{\pi}{-\ln x}} \leq \sum_{n=0}^{\infty} x^{n^2} \leq 1 + \frac{1}{2}\sqrt{\frac{\pi}{-\ln x}},$$

so that

$$\sum_{n=0}^{\infty} x^{n^2} \sim \frac{1}{2}\sqrt{\frac{\pi}{1-x}}$$

as $x \to 1^-$.

18.2 Choose A so that $|x_n| < A$ for all n. For any $\epsilon > 0$, choose n_1 so that $|x_n - x| < \epsilon/2$ for $n \geq n_1$. Then choose n_2 so that $n_1 \cdot (A + |x|)/n_2 < \epsilon/2$. Let $N = \max\{n_1, n_2\}$. For $n \geq N$, we have

$$
\begin{aligned}
|y_n - x| &= \frac{|x_1 + \cdots + x_n - nx|}{n} \\[2mm]
&\leq \frac{|x_1 - x| + \cdots + |x_{n_1} - x|}{n} + \frac{|x_{n_1+1} - x| + \cdots + |x_n - x|}{n} \\[2mm]
&\leq \frac{n_1 \cdot (A + |x|)}{n_2} + \frac{n - n_1}{n} \cdot \frac{\epsilon}{2} \\[2mm]
&< \frac{\epsilon}{2} + \frac{\epsilon}{2} \\[2mm]
&= \epsilon,
\end{aligned}
$$

and hence $\lim_{n \to \infty} y_n = x$.

18.3 Let $S_n = \sum_{k=1}^{n} a_k/k$. We are given that $S_n \to A$ for some A. By Abels' formula, we have

$$\sum_{k=1}^{n} a_k = \sum_{k=1}^{n}\left(\frac{1}{k}a_k\right) \cdot k = S_n \cdot n - \sum_{j=1}^{n-1} S_j.$$

Therefore, by Problem 18.2,

$$\frac{1}{n}\sum_{k=1}^{n} a_k = S_n - \frac{1}{n}\sum_{j=1}^{n-1} S_j$$

$$= S_n - \frac{n-1}{n} \cdot \left(\frac{1}{n-1} \cdot \sum_{j=1}^{n-1} S_j\right)$$

$$\rightarrow A - 1 \cdot A = 0.$$

Additional Problems

18.4 Define a family of curves by

$$S_n = \left\{(x, y) : y = \frac{1}{n}\sin(n^2 x), 0 \le x \le \pi\right\},$$

where n is a positive integer. What is the limit of the length of S_n as $n \to \infty$?

18.5 Show that

$$\sum_{n=1}^{\infty}(x + 2^n)^{-1},$$

as a function of x, goes to 0 as $x \to \infty$. Estimate how fast it does so.

18.6 Given a, r, b with $a < r < b$, show that

$$\int_a^b e^{-c(x-r)^2}\, dx \sim \left(\frac{\pi}{c}\right)^{1/2}$$

as $c \to \infty$.

18.7 Suppose that the series $\sum_{n=1}^{\infty}(\pm 1)/n$ is convergent, where the signs are prescribed. Show that the number of +1's up to a given point is asymptotic to the number of −1's up to that point.

18.8 Let

$$f(n) = 1 \cdot 2^2 \cdot 3^3 \cdot \cdots \cdot n^n.$$

Estimate $f(n)$.

18.9 Show that there is a constant $c > 0$ such that

$$\left|\int_x^{\infty} e^{-u^2}\, du - \frac{e^{-x^2}}{2x}\right| < \frac{ce^{-x^2}}{x^3}$$

for $x > 0$.

18.10 Show that for every real number x and every positive integer N,

$$\left|\sum_{n=1}^{N} \frac{\sin 2\pi n x}{n}\right| < 1.$$

18.11 Suppose that f'' is continuous and nonnegative, and that $f'(a) > 0$.
Show that

$$\left| \int_a^b \cos(f(x)) \, dx \right| \le \frac{2}{f'(a)}.$$

18.12 Show that

$$\sum_{k=0}^{n} \binom{n}{k}^2 \binom{n+k}{k}^2 = (1 + \sqrt{2})^{4n(1+o(1))}.$$

18.13 Prove that

$$\frac{2^{2n}}{2\sqrt{n}} < \binom{2n}{n} < \frac{2^{2n}}{\sqrt{2n}}.$$

18.14 Let $L(R)$ denote the number of lattice points in the disc $x^2 + y^2 \le R^2$. Show that $|L(R) - \pi R^2| \le cR$, for some constant c.

18.15 Let $P(x)$ be the polynomial

$$P(x) = x^n + a_1 \binom{n}{1} x^{n-1} + \cdots + a_{n-1} \binom{n}{n-1} x + a_n.$$

Let $A = \max_{1 \le k \le n} |a_k|^{1/k}$, and let M be the modulus of the largest zero of $P(x)$. Show that $A \le M \le A/(2^{1/n} - 1)$.

18.16 (*Math. Magazine*, Problem 1310, December 1988) Show that

$$\sum_{n=1}^{\infty} 1/(\mathrm{lcm}\{1, 2, 3, \ldots, n\})^\delta$$

converges for all $\delta > 0$. ($\mathrm{lcm}\{1, 2, 3, \ldots, n\}$ is the least common multiple of the set $\{1, 2, 3, \ldots, n\}$.)

Chapter 19

Deus Ex Machina

> Do you promise that your Detectives shall well and truly detect the Crimes presented to them, using those Wits which it shall please you to bestow upon them and not placing reliance upon, nor making use of, Divine Revelation, Feminine Intuition, Mumbo-Jumbo, Jiggery-Pokery, Coincidence or the Act of God?
>
> DOROTHY SAYERS
> *The Mind of the Maker*, 1941

No, we don't promise that. In this chapter, the solutions require elements not explicit in the problem statements. These new elements—the "extras"—are often surprising. Deciding what elements to introduce is a real test of mathematical creativity, and that is why we reserve these problems for the second-to-last chapter.

Example 19.1. Seven particles travel back and forth in the unit interval $[0,1]$. Initially they all move to the right with the same speed. When a particle strikes another particle, or reaches the numbers 0 or 1, its direction of motion changes (while its speed remains constant). How many particle–particle collisions occur before the particles again occupy their original positions and are moving to the right?

Solution: Suppose that there is only one particle moving back and forth in the unit interval. Figure 19.1(a) shows the path of the particle over time. Superimposing seven such graphs, we obtain the paths of all seven particles, as in Figure 19.1(b). For the purposes of the diagram, when two particles collide, we can just as well assume that they pass through each other. If their time-paths intersect, then this counts as a collision. It is easy to see in Figure 19.1(b) that there are $2\binom{7}{2} = 42$ intersections. ∎

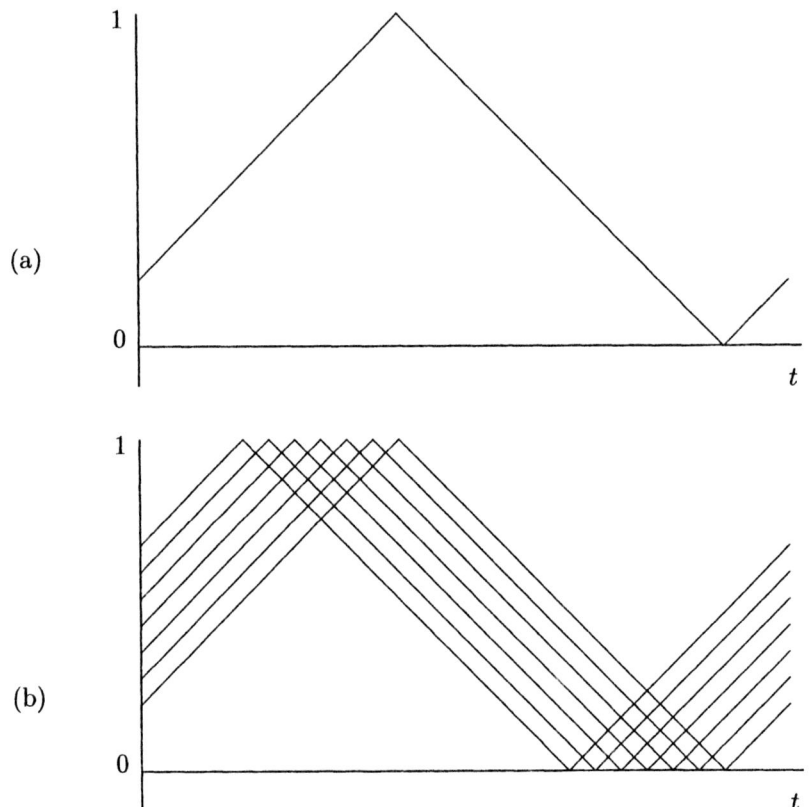

(a)

(b)

Figure 19.1: (a) The path of one particle, and (b) the paths of seven colliding particles.

Example 19.2. Evaluate the integral

$$I = \int_{-\infty}^{\infty} e^{-x^2} \, dx.$$

Solution: The usual substitutions and other methods don't seem to work. The "a ha!" technique is to compute a related double integral. Thus

$$
\begin{aligned}
I^2 &= \int_{-\infty}^{\infty} e^{-x^2} \, dx \int_{-\infty}^{\infty} e^{-y^2} \, dy \\
&= \int_{-\infty}^{\infty} \int_{-\infty}^{\infty} e^{-x^2 - y^2} \, dx \, dy.
\end{aligned}
$$

We introduce polar coordinates r and θ, $0 \le r < \infty$, $0 \le \theta \le 2\pi$, with $x^2 + y^2 = r^2$ and $dx\, dy = r\, dr\, d\theta$. Now

$$I^2 = \int_0^{2\pi} \int_0^\infty e^{-r^2} r\, dr\, d\theta$$

$$= \int_0^{2\pi} d\theta \int_0^\infty re^{-r^2}\, dr$$

$$= 2\pi \left(-\frac{1}{2} e^{-r^2} \right)\Big]_0^\infty$$

$$= \pi,$$

and

$$I = \sqrt{\pi}.$$

Note: By a change of variables, we can write our result in the more general form,

$$\frac{1}{\sqrt{2\pi}\,\sigma} \int_{-\infty}^\infty e^{-\frac{1}{2}\left(\frac{x-\mu}{\sigma}\right)^2}\, dx = 1,$$

which represents the total distribution of a normal random variable with mean μ and standard deviation σ. ∎

Example 19.3. Find all integers x, y with $x^3 + y^3 = 6xy$.

Solution: Note that

$$a^3 + b^3 + c^3 - 3abc = (a + b + c)(a^2 + b^2 + c^2 - ab - bc - ac).$$

Letting $a = x$, $b = y$, and $c = 2$, we obtain

$$(x + y + 2)(x^2 + y^2 + 4 - xy - 2x - 2y) = x^3 + y^3 + 8 - 6xy = 8.$$

The integer factorizations of 8 are $1 \cdot 8$, $2 \cdot 4$, $4 \cdot 2$, $8 \cdot 1$, $-1 \cdot -8$, $-2 \cdot -4$, $-4 \cdot -2$, and $-8 \cdot -1$. It is now easy to show that $(0,0)$ and $(3,3)$ are the only solutions to the given equation. If $x + y + 2 = 8$, then $x + y = 6$ and hence $36 - 12 + 4 - 3xy = 1$, $xy = 9$, which gives the solution $x = 3$, $y = 3$. The case $x + y + 2 = 2$ produces $x = 0$, $y = 0$. The other cases fail to give solutions. For example, if $x + y + 2 = -4$, we get the equations $x + y = -6$ and $xy = 18$, which have no common integer solutions. Hence the only integer solutions are $(x, y) = (0, 0)$ and $(3, 3)$. ∎

Problems

19.1 (Putnam Competition, 1984) Find the minimum value of

$$(u - v)^2 + \left(\sqrt{2 - u^2} - \frac{9}{v}\right)^2$$

for $0 < u < \sqrt{2}$ and $v > 0$.

19.2 Suppose that the unit disc is covered with n infinitely long rectangular strips of widths w_i, $1 \leq i \leq n$. Prove that the sum of the w_i is at least 2.

19.3 Two sets are said to be *almost disjoint* if their intersection is finite. Exhibit uncountably many sets of positive integers each pair of which is almost disjoint.

19.4 A *chain* is a totally ordered subset of a partially ordered set. Prove or disprove: There exists an uncountable chain of subsets of **N** (where subsets are ordered by set inclusion).

19.5 Given three nonoverlapping circles (of different radii) in the plane, show that the three pairs of of common external tangents of the circles intersect in three collinear points.

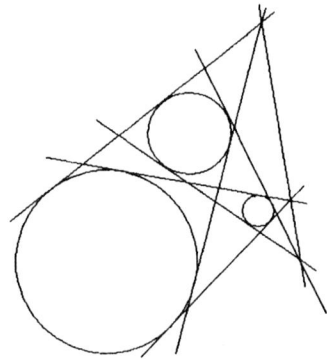

Solutions

19.1 Let $f(u,v) = (u-v)^2 + (\sqrt{2-u^2} - 9/v)^2$. Then $f(u,v)$ is the square of the distance between the ordered pairs $(u, \sqrt{2-u^2})$ and $(v, 9/v)$. These points lie, respectively, on the circle $x^2 + y^2 = 2$ and the hyperbola $xy = 9$. It is clear that two such points at minimum distance also lie on the line $y = x$. Hence $u = \sqrt{2-u^2}$ and $v = 9/v$. The minimum value of $f(u,v)$ is therefore $f(1,3) = 8$.

19.2 Let S be the unit sphere with the same center as the given circle. Project each strip onto S. We claim that the area of the projection of the ith strip is $2\pi w_i$. To see this, we compute the surface area obtained by rotating that portion of the strip which lies inside the unit circle around an axis perpendicular to it. We obtain:

$$A = 2\pi \int_a^b y\sqrt{1 + \frac{x^2}{y^2}}\, dx = 2\pi(b-a) = 2\pi w_i.$$

Since the circle is covered by the strips, the surface of the sphere is also covered. Therefore $\sum_{i=1}^n 2\pi w_i \geq 4\pi$ and $\sum_{i=1}^n w_i \geq 2$.

19.3 Solution (1): Thinking geometrically, we formulate the sets of positive integers based on the lines $y = mx$ in the plane, where m is a real number. Let f be a bijection between $\mathbf{N}$ and the set of lattice points in the plane (points with both coordinates integers). For each line $y = mx$, define a subset of integers A_m as follows: for each positive integer i, let A_m contain the value $f(p)$, where p is the lattice point closest to the intersection of the lines $y = mx$ and $y = i$. It is clear that there are uncountably many lines and that each pair of lines "deviate" from each other so that the corresponding sets of integers are almost disjoint.

Solution (2): For each real number t, let A_t be the set of decimal digits of better and better approximations to the number t.

19.4 We will prove the assertion. Let f be a bijection between $\mathbf{Q}$ and $\mathbf{N}$. For each real number t, let A_t be the set of rational numbers q for which $q \leq t$. Then the images $f(A_t)$ are an uncountable chain of sets of integers.

19.5 Let the given circles be C_1, C_2, and C_3, and suppose that they lie in the plane π_1. Let S_1, S_2, S_3 be the spheres that have C_1, C_2, C_3, respectively, as great circles. The centers of these spheres lie in π_1. Let π_2 be a plane tangent to S_1, S_2, and S_3 (there are two such planes; choose either). Then π_2 is also tangent to three cones that are tangent to pairs of spheres. Therefore, π_2 contains the vertices of the cones, namely, the three intersection points of the external tangents. Hence the intersection points lie in the intersection of two planes and are therefore collinear.

Additional Problems

19.6 For what real numbers r does the system of equations

$$x^2 = y^2$$
$$(x - r)^2 + y^2 = 1$$

have exactly one, two, three, and four solutions?

19.7 Find formulas for $\sin^n x$ and $\cos^n x$ in terms of $\sin x$ and $\cos x$.

19.8 (*Amer. Math. Monthly*, Problem 10447, 1995; modified) Suppose that we are given a finite tournament on a finite set of teams. Let L_0 be a listing of the teams in some order, and define successively L_i, $i = 1$, $2, 3, \ldots$ by repeated application of the following operation: if a team in the list L_i lost to the team immediately following it in the list, call that pair of teams a *switchable pair*; the order of one switchable pair is then reversed to give L_{i+1}. Note that this may increase the number of switchable pairs.

Prove that any such sequence of operations leads, in a finite number of steps, to a list in which every team defeated the team immediately following it in the list, so there are no switchable pairs.

19.9 Find, for $n \geq 1$, a formula for

$$I_n = \int_0^\pi \left(\frac{\sin nx}{\sin x} \right)^2 dx.$$

19.10 (J. H. Conway) Suppose that a triangle has sides of integral length and angles that are integral multiples of $2\pi/n$ for some n. Prove that the triangle is equilateral.

19.11 Suppose that a set of integers with positive sum is arranged in a circle. A "move" consists of replacing three consecutive terms in the circle, x, y, z, where y is in the middle and $y < 0$, with the terms $x + y$, $-y$, $z + y$. Show that one may make only a finite number of moves.

The picture below illustrates a move based on the numbers 1, -1, -3.

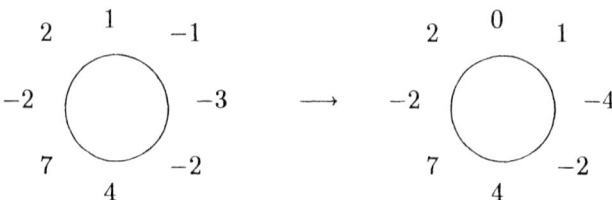

Hint: Assign a particular positive integer-valued function f to the set of circular sequences of integers and show that that f decreases as a result of every move.

19.12 Let p be a prime number greater than 3. As $S = \{1,\ldots,p-1\}$ is a group under multiplication modulo p, for every k there exists a unique $x_k \in S$ such that $kx_k \equiv 1 \bmod p$. Thus integers $n_1, \ldots, n_{p-1}$ are specified such that

$$kx_k = 1 + n_k p$$

for $k = 1, \ldots, p-1$. Show that

$$\sum_{k=1}^{p-1} k n_k \equiv \frac{1}{2}(p-1) \bmod p.$$

19.13 Can a finite number of unit squares be placed in the plane so that every corner of every square is a corner of exactly one other square?

19.14 Evaluate the integral

$$\int_0^\infty \frac{\sin nx \cos^n x}{x}\, dx,$$

where n is a positive integer.

Chapter 20

More Problems

"You know my methods. Apply them."

SIR ARTHUR CONAN DOYLE
Sherlock Holmes, *The Sign of Four*, 1890

Now that we have practiced various problem solving methods, we need to try them on some problems not organized by topic. An immediate question arises: What technique should we use on each problem? This will be the main consideration when tackling the following exercises.

Although there is no general algorithm (that we know of) for solving any and all problems that come your way, there are definitely some things you can do to become a better problem solver. The following advice is necessarily very general:

1. Solve lots of problems. We learn by doing. Even if you don't have immediate success you will be improving your skills.

2. Keep a record of your solutions. Perhaps a new problem can be solved using an old technique.

3. Try to solve the problems yourself. If necessary, look up a solution or ask someone else. But once you learn the answer, think through all the steps carefully on your own.

4. Remember and use the most general principles of problem solving:

    (a) If the problem calls for a formula or unknown value, try generating some data and making a guess as to the general pattern. Try proving your conjecture by induction.

(b) If the problem is a statement to be proved, try proving a special case first. Perhaps the proof of the special case will give you insight into the proof of the more general statement. Going in the other direction, perhaps some parts of the hypothesis are unnecessary. Does the result follow under less stringent conditions? By eliminating the extraneous it may be easier to concentrate on the relevant.

(c) Use symmetry to simplify the problem.

(d) If the problem contains messy calculations involving integers, try a modulo n argument with a modulus that seems suited to the problem, and see if the computations are made easier.

(e) If the problem asks for a proof of a statement of the type, "There exists a ..." or "There exist at least fifteen ...", try using the pigeonhole principle.

(f) If the problem asks for a proof of an identity, try using the method of counting something two different ways.

5. If you can't solve a problem and you can't find a solution elsewhere, just put the problem at the back of your mind for a time. Perhaps an inspiration will come later.

6. Our final word: Don't give up. The most rewarding problems are often the ones that appear the most difficult at first. Perhaps it is fitting to mention here a problem that the authors do not know how to solve. We would be happy to hear from anyone who solves it.

Open Problem. Show that $n^2 + 1$ unit squares cannot be arranged in the plane so as to cover a square of side length greater than n. (The perimeter and interior of the large square must be covered.)

Now you are on your own for a while. You must decide what methods to use to solve the problems of this chapter. As usual, solutions follow, and following the solutions is an additional set of problems to further challenge you. Good luck!

Problems

20.1 The unit of exchange on the planet Pluto is the Pluton, and currency exists only in denominations of 5 and 7 Plutons. What integer amounts can Plutonians pay without receiving change?

20.2 Suppose that P is the convex hull of a finite number of points in $\mathbf{R}^n$ and that f is a real-valued convex function on P. Prove that f has a maximum value on P that occurs at some vertex of P.

20.3 (Putnam Competition, 1991) Suppose p is an odd prime. Prove that

$$\sum_{j=0}^{p} \binom{p}{j}\binom{p+j}{j} \equiv 2^p + 1 \bmod p^2.$$

20.4 Find nonconstant rational functions x, y of t such that

$$y^2 + 2y = x^3 - x^2 - x$$

identically in t.

20.5 An urn contains balls numbered $1, \ldots, n$ but otherwise identical. A ball is picked from the urn, its number noted, and then returned to the urn. This operation is performed three times. Prove that the probability that the sum of the three numbers obtained is divisible by 3 is at least $1/4$.

20.6 (Putnam Competition, 1990) If A and B are square matrices of the same size such that $ABAB = 0$, does it follows that $BABA = 0$?

20.7 Find a_1, a_2, a_3, a_4, a_5, a_6 such that

$$f(x) = 1 + a_1 e^x + a_2 e^{2x} + a_3 e^{3x} + a_4 e^{4x} + a_5 e^{5x} + a_6 e^{6x}$$

has a zero of order exactly six at $x = 0$.

20.8 Prove the identity

$$1 = \sum_{k=1}^{n} \prod_{\substack{j=1 \\ j \neq k}}^{n} \left(1 - \frac{x_k}{x_j}\right)^{-1}.$$

20.9 Let x be any number of the form

$$\pm\sqrt{2 \pm \sqrt{2 \pm \cdots \pm \sqrt{2}}},$$

where there are n square root signs. Prove that $(2^{n+1}/\pi)\cos^{-1}(x/2)$ is an odd integer.

20.10 If $f(x) = \tan x$, prove that $f^{(n)}(0) \geq 0$ for all $n \geq 0$.

20.11 Find functions $f(x)$ and $g(x)$ on $\mathbf{R}$ such that $f(g(x))$ is, but $g(f(x))$ is not, identically equal to x.

20.12 Find a polynomial P in six variables such that $P(x_1, y_1, x_2, y_2, x_3, y_3)$ vanishes if and only if the points (x_1, y_1), (x_2, y_2), (x_3, y_3) are the vertices of an equilateral triangle.

20.13 Let $P(z)$ be a nonconstant polynomial and let z be a complex variable. Show that the roots of the polynomial $P'(z)$ are contained in the convex hull of the roots of $P(z)$.

20.14 Let $a = e^{-e}$. Suppose that x_0 is an arbitrary real number and $x_{n+1} = a^{x_n}$ for $n = 0, 1, \ldots$. Prove that the sequence x_0, x_1, x_2, ...has a limit and determine its value.

20.15 For $n \geq 2$, let $x_1, \ldots, x_n$ be nonzero real numbers whose sum is zero. Show that there are i, j with $1 \leq i < j \leq n$ such that

$$1/2 \leq |x_i/x_j| \leq 2.$$

20.16 Find a nonconstant polynomial P with rational coefficients such that $Q = P^2 - 2$ has a nontrivial polynomial factor also with rational coefficients (i.e., a factor whose degree is strictly between 0 and the degree of Q).

20.17 Let k be a fixed number greater than 1. Define a function f on the interval $(0, \pi/2k)$ by the formula $f(x) = \sin kx / \sin x$. Prove that f has a negative derivative on that interval.

20.18 For $n \geq 1$ let $a_1, \ldots, a_n$ and $b_1, \ldots, b_n$ be complex numbers satisfying

$$a_1^r + \cdots + a_n^r = b_1^r + \cdots + b_n^r$$

for all integers $r \geq 2$. Prove that this equation also holds for $r = 1$.

20.19 Let $a_1 = 1$ and $a_{n+1} = n/a_n$ for $n \geq 1$. Find

$$\lim_{n \to \infty} n^{-1/2}(1/a_1 + \cdots + 1/a_n).$$

Solutions

20.1 We generate some data and find that the possible integer amounts that Plutonians can pay without receiving change are

$$5, 7, 10, 12, 14, 15, 17, 19, 20, 21, 22, 24, 25, 26, 27, 28, \ldots.$$

In addition to some sporadic small values, it appears that this list consists of all integers greater than or equal to 24. We prove this by induction. Suppose that an amount $n \geq 24$ can be paid. Then $n = 5a + 7b$ for some nonnegative a and b. Note that $a \geq 4$ or $b \geq 2$, or else $n \leq 22$. Now $n + 1 = 5(a + 3) + 7(b - 2)$ and also $n + 1 = 5(a - 4) + 7(b + 3)$. Thus $n + 1$ can be paid. By induction, all amounts $n \geq 24$ can be paid.

20.2 Let $v_1, \ldots, v_n$ be the vertices of P and let

$$M = \max\{f(v_i) : 1 \leq i \leq n\}.$$

Each x in P can be expressed in the form $x = \sum_{i=1}^{n} t_i v_i$, where each $t_i \geq 0$ and $\sum_{i=1}^{n} t_i = 1$. Then

$$f(x) = f\left(\sum_{i=1}^{n} t_i v_i\right) \leq \sum_{i=1}^{n} t_i f(v_i) \leq M \cdot \sum_{i=1}^{n} t_i = M.$$

20.3 By Vandermonde's identity (Glossary),

$$\sum_{j=0}^{p} \binom{p}{j}\binom{p+j}{j} = \sum_{j=0}^{p} \binom{p}{j} \sum_{k=0}^{j} \binom{j}{k}\binom{p}{j-k}$$

$$= \sum_{0 \leq k \leq j \leq p} \binom{p}{j}\binom{j}{k}\binom{p}{j-k}.$$

Note that $\binom{p}{j}$ and $\binom{p}{j-k}$ are each divisible by p unless they equal 1, so each summand is divisible by p^2 except when $j = p$ and $k = 0$ or $j = k$. Therefore,

$$\sum_{j=0}^{p} \binom{p}{j}\binom{p+j}{j} \equiv \binom{p}{p}\binom{p}{0}\binom{p}{p} + \sum_{k=1}^{p} \binom{p}{k}\binom{k}{k}\binom{p}{0}$$

$$\equiv 1 + \sum_{k=1}^{p} \binom{p}{k}$$

$$\equiv 1 + 2^p \bmod p^2.$$

20.4 From $y^2 + 2y = x^3 - x^2 - x$, it follows that $y^2 + 2y + 1 = x^3 - x^2 - x + 1$, and hence $(y + 1)^2 = (x + 1)(x - 1)^2$. Let $x = s^2 - 1$. Then $(y + 1)^2 = s^2(s^2 - 2)^2$, so that $y + 1 = \pm(s^2 - 2)s$. Thus $(x, y) = (s^2 - 1, \pm(s^2 - 2)s - 1)$ is a solution of the desired kind when s is any rational function of t.

20.5 Let P_n be the probability that, when three balls are chosen, the sum of their numbers is divisible by 3. Note that $P_1 = 1$ and $P_2 = 1/4$. We must show that $P_n \geq \frac{1}{4}$ for all $n \geq 3$. The only relevant thing about the integers $1, \ldots, n$ is their residues modulo 3. Thus, when $n \geq 3$, we may think of there being three integers, $0, 1, 2$, occuring in the urn with certain probabilities. We consider three cases:

(1) $n = 3k$. Here $\Pr(i = 0) = k/n$, $\Pr(i = 1) = k/n$, and $\Pr(i = 2) = k/n$. The only triples of residues that sum to 0 modulo 3 are 000, 111, 222, and 012. Thus

$$
\begin{aligned}
P_n &= \frac{3k^3 + 3k^3 + 3k^3 + 6k^3}{n^3} \\
&= \frac{9k^3}{n^3} \\
&= \frac{1}{3}.
\end{aligned}
$$

(2) $n = 3k + 1$. Here $\Pr(i = 0) = k/n$, $\Pr(i = 1) = (k+1)/n$, and $\Pr(i = 2) = k/n$.

$$
\begin{aligned}
P_n &= \frac{k^3 + (k+1)^3 + k^3 + 6k^2(k+1)}{n^3} \\
&= \frac{9k^3 + 9k^2 + 3k + 1}{27k^3 + 27k^2 + 9k + 1} \\
&> \frac{1}{3}.
\end{aligned}
$$

(3) $n = 3k + 2$. Here $\Pr(i = 0) = k/n$, $\Pr(i = 1) = (k+1)/n$, and $\Pr(i = 2) = (k+1)/n$. Thus

$$
\begin{aligned}
P_n &= \frac{k^3 + (k+1)^3 + (k+1)^3 + 6k(k+1)^2}{n^3} \\
&= \frac{9k^3 + 18k^2 + 12k + 2}{27k^3 + 54k^2 + 18k + 8} \\
&= \frac{1}{3} - \frac{2}{3(27k^3 + 54k^2 + 18k + 8)} \\
&\geq \frac{1}{4}.
\end{aligned}
$$

20.6 No. Let

$$A = \begin{bmatrix} 1 & 0 & 0 \\ 0 & 1 & 0 \\ 0 & 0 & 0 \end{bmatrix}$$

and

$$B = \begin{bmatrix} 0 & 0 & 0 \\ 1 & 0 & 0 \\ 0 & 1 & 0 \end{bmatrix}.$$

Direct calculation shows that $ABAB = 0$ and $BABA \neq 0$.

Note: We may think of A acting on three components x, y, z. A has no effect on x and y but acts as a "trapdoor" on z. The action of B is to change y to x, change z to y, and to act as a trapdoor for x.

20.7 Let $z = e^x$, so that

$$g(z) = 1 + a_1 z + a_2 z^2 + a_3 z^3 + a_4 z^4 + a_5 z^5 + a_6 z^6.$$

Since we need $z = 1$ to be a zero of order 6 of this polynomial, it follows that

$$g(z) = (1 - z)^6.$$

Therefore

$$g(z) = 1 - 6z + 15z^2 - 20z^3 + 15z^4 - 6z^5 + z^6.$$

Thus the coefficients of f are -6, 15, -20, 15, -6, 1.

20.8 Let $P(x) = (x - x_1)\dots(x - x_n)$. Then

$$\frac{1}{P(x)} = \frac{a_1}{x - x_1} + \dots + \frac{a_n}{x - x_n},$$

for some constants $a_1, \dots, a_n$. Now

$$a_1 = \frac{1}{(x_1 - x_2)\dots(x_1 - x_n)} = \lim_{x \to x_1} \frac{x - x_1}{P(x)} = \frac{1}{P'(x_1)}.$$

Thus $a_k = \prod_{j \neq k}\left(\frac{1}{x_k - x_j}\right)$ and

$$\frac{1}{\prod_{l=1}^{n}(x - x_l)} = \sum_{k=1}^{n} \frac{1}{x - x_k} \prod_{\substack{j=1 \\ j \neq k}}^{n}\left(\frac{x_k}{x_j} - 1\right)^{-1} x_j$$

$$= \left(\prod_{j=1}^{n} x_j^{-1}\right) \sum_{k=1}^{n} \frac{x_k}{x - x_k} \prod_{\substack{j=1 \\ j \neq k}}^{n}\left(\frac{x_k}{x_j} - 1\right)^{-1}.$$

Evaluating this expression at $x = 0$ gives the desired identity.

20.9 The proof is by induction on n. For $n = 1$, $x = \pm\sqrt{2}$ so $\cos^{-1}\frac{x}{2}$ is either $\frac{\pi}{4}$ or $\frac{3\pi}{4}$ and we have $\frac{4}{\pi} \cdot \frac{\pi}{4} = 1$ and $\frac{4}{\pi} \cdot \frac{3\pi}{4} = 3$.

Now suppose that the result is true for some $n \geq 1$ and let

$$x = \pm\sqrt{2 \pm \sqrt{2 \pm \cdots \pm \sqrt{2}}}$$

with $n + 1$ square roots. Then

$$x^2 - 2 = \pm\sqrt{2 \pm \sqrt{2 \pm \cdots \pm \sqrt{2}}}$$

with n square roots, so by assumption,

$$\frac{2^{n+1}}{\pi} \cos^{-1}\left(\frac{x^2 - 2}{2}\right) = k \tag{20.1}$$

where k is an odd integer. By the double angle formula,

$$\cos\left(2\cos^{-1}\frac{x}{2}\right) = 2 \cdot \frac{x^2}{4} - 1 = \frac{x^2 - 2}{2}.$$

This means that either $\cos^{-1}\frac{x^2-2}{2} = 2\cos^{-1}\frac{x}{2}$ or $\cos^{-1}\frac{x^2-2}{2} = 2\pi - 2\cos^{-1}\frac{x}{2}$. Substitution of either of these into the equation (20.1) shows that $\frac{2^{n+2}}{\pi}\cos^{-1}\frac{x}{2}$ is an odd integer.

20.10 Obviously, $f^{(0)} = \tan 0 = 0$. Next we observe that $f'(x) = \sec^2 x = 1+\tan^2 x$, so $f'(0) = 1$. The result will follow by using math induction to show that $f^{(n)}(x)$ is always a polynomial in $\tan x$ (of degree $n+1$) with all nonnegative coefficients. This is clear for $n = 1$. If for some n we have

$$f^{(n)}(x) = \sum_{j=0}^{n+1} a_j \tan^j(x)$$

with each $a_j \geq 0$, then

$$f^{(n+1)}(x) = \sum_{j=1}^{n+1} ja_j \tan^{j-1}(x)\sec^2 x = \sum_{j=1}^{n+1} ja_j \tan^{j-1}(x)(1 + \tan^2 x),$$

a polynomial in $\tan x$ of degree $n+2$, also having nonnegative coefficients.

Note: In fact, $f^{(n)}(0) = 0$ for all even n.

20.11 Let $f(x) = \ln|x|$ if $x \neq 0$ and 0 if $x = 0$. Let $g(x) = e^x$. Then

$$f(g(x)) = \ln|e^x| = \ln e^x = x$$

but $g(f(x)) \neq x$; for example,

$$g(f(-2)) = e^{f(-2)} = e^{\ln 2} = 2 \neq -2.$$

20.12 Let $A = (x_1, y_1)$, $B = (x_2, y_2)$, $C = (x_3, y_3)$ and take

$$P = (|AB|^2 - |AC|^2)^2 + (|AB|^2 - |BC|^2)^2.$$

This is a polynomial of degree four.

20.13 Assume that $P(z) = A \cdot \prod_{j=1}^{n}(z - r_j)$, where $r_1, \ldots, r_n$ are the zeros of $P(z)$. Then we see formally that

$$\frac{P'(z)}{P(z)} = \sum_{j=1}^{n} \frac{1}{z - r_j}.$$

Note that if $P'(z) = 0$ and $P(z) = 0$ then the problem result is trivially true. So assume that $P'(z) = 0$ and $P(z) \neq 0$. Then we have

$$\sum_{j=1}^{n} \frac{1}{z - r_j} = 0$$

or equivalently

$$\sum_{j=1}^{n} \frac{z - r_j}{|z - r_j|^2} = 0.$$

Hence

$$\sum_{j=1}^{n} \frac{z - r_j}{|z - r_j|^2} = 0,$$

which may be written

$$z \cdot \alpha = \sum_{j=1}^{n} \frac{r_j}{|z - r_j|^2}$$

where

$$\alpha = \sum_{j=1}^{n} \frac{1}{|z - r_j|^2}.$$

Therefore

$$z = \sum_{j=1}^{n} \frac{r_j}{\alpha|z - r_j|^2}.$$

and, for each j,

$$\frac{1}{\alpha|z - r_j|^2} > 0,$$

and

$$\sum_{j=1}^{n} \frac{1}{\alpha|z - r_j|^2} = \frac{1}{\alpha} \cdot \alpha = 1.$$

20.14 We will show that $\lim_{n \to \infty} x_n = 1/e$ by first showing that $\{x_{2n}\}$ and $\{x_{2n+1}\}$ both converge, and then showing that each has limit $1/e$.

Note that $f(x) = a^x$ is decreasing and $g(x) = a^{a^x}$ is increasing on $\mathbf{R}$. Also, g is bounded on $\mathbf{R}$ since $\lim_{x \to \infty} g(x) = 1$ and $\lim_{x \to -\infty} g(x) = 0$.

Suppose that $x_0 \le x_2$. By induction, $\{x_{2n}\}$ is increasing since if $x_{2k} \le x_{2k+2}$ then

$$x_{2k+2} = a^{a^{x_{2k}}} \le a^{a^{x_{2k+2}}} = x_{2k+4}.$$

Also,

$$x_{2n} = a^{a^{x_{2n-2}}} = g(x_{2n-2}) < 1,$$

so $\{x_{2n}\}$ is bounded above and hence convergent. The sequence $\{x_{2n+1}\}$ is also convergent since

$$x_{2n+1} = a^{x_{2n}} \ge a^{x_{2n+2}} = x_{2n+3},$$

i.e., $\{x_{2n+1}\}$ is decreasing (and bounded above).

For the case $x_0 > x_2$, a similar argument shows that $\{x_{2n}\}$ is decreasing and $\{x_{2n+1}\}$ is increasing.

So assume that $x_{2n} \to A$ and $x_{2n+1} \to B$. From the recurrence relation we get $A = a^B$ and $B = a^A$, and hence $A = a^{a^A}$ and $B = a^{a^B}$. Since $a^{a^{1/e}} = 1/e$, the result will follow if we can show that the function $h(x) = a^{a^x} - x$ is one-to-one.

By differentiating we obtain

$$h'(x) = e^2 a^{a^x + x} - 1$$

and

$$h''(x) = -e^3 a^{a^x + x}[1 - ea^x].$$

Note that $h'(1/e) = 0$ and that $x = 1/e$ is the only solution of $h''(x) = 0$. It is straightforward to check that $h''(0) > 0$ and $h''(1) < 0$, so by the first derivative test (applied to h') we can conclude that $h'(x) < 0$ for $x \ne 1/e$ and hence h is strictly decreasing and thus h is one-to-one.

20.15 Solution (1): Let $a = \min\{x_i\}$ and $b = \max\{x_i\}$. We may assume (by replacing each x_i with $-x_i$ if necessary) that $|a| > b$. If $|a| \leq 2b$, then $1 \leq |a/b| \leq 2$ and we are done. So assume that $b < |a|/2$.

Now consider the (infinitely many) intervals

$$\cdots, \left[\frac{|a|}{2^n}, \frac{|a|}{2^{n-1}}\right), \cdots, \left[\frac{|a|}{8}, \frac{|a|}{4}\right), \left[\frac{|a|}{4}, \frac{|a|}{2}\right).$$

One of these intervals must contain two of the positive x_i values, for otherwise

$$\sum_{x_i>0} x_i < \frac{|a|}{2} + \frac{|a|}{4} + \cdots = |a| \cdot \left(\frac{1}{2} + \frac{1}{4} + \frac{1}{8} + \cdots\right) = |a|,$$

which is impossible since $\sum_{i=1}^{n} x_i = 0$. So there exist $i \neq j$ such that $x_i, x_j \in \left[\frac{|a|}{2^n}, \frac{|a|}{2^{n-1}}\right)$ for some n and consequently $1/2 < x_i/x_j < 2$.

Solution (2): Suppose that there is no such pair x_i, x_j. Then for each $m \in \mathbf{Z}$, there is at most one n for which $2^{m-1} \leq |x_n| < 2^i$. Let j be such that $|x_j|$ is greatest and suppose that $2^{m-1} \leq |x_j| < 2^m$. Then $|x_k| < 2^{m-1}$ for $k \neq j$. But

$$\sum_{k \neq j} |x_k| < \sum_{-\infty < l < m-1} 2^l = 2^{m-1}.$$

Hence

$$-\sum_{k \neq j} |x_k| + |x_j| \neq 0,$$

so that even if all the x_k had opposite sign of x_m we would not have $\sum_{i=1}^{n} x_i = 0$.

20.16 Starting with the form $P(x) = x^2 + a$, then matching coefficients in the equation

$$P(x)^2 - 2 = (x^2 + bx + c)(x^2 + dx + e),$$

we obtain $P(x) = x^2 - \frac{3}{2}$. With this choice of P, we have $Q(x) = P(x)^2 - 2 = x^4 - 3x^2 + \frac{1}{4}$, which factors as

$$Q(x) = \left(x^2 + 2x + \frac{1}{2}\right)\left(x^2 - 2x + \frac{1}{2}\right).$$

20.17 Examining $f'(x)$, we see that $f'(x) < 0$ provided $k \sin x \cos kx - \cos x \sin kx = \cos x \cos kx(k \tan x - \tan kx) < 0$. Since x and kx are in $(0, \pi/2)$, the result follows provided $\tan kx - k \tan x > 0$. To show this, let $g(x) = \tan kx - k \tan x$. Then $g(0) = 0$ and $g'(x) = k(\sec^2 kx - \sec^2 x) > 0$ since the function $\sec^2 x$ is strictly increasing on $(0, \pi/2)$. It follows that $g(x) > 0$ on $(0, \pi/2)$.

20.18 Note that the result is trivial if all $a_i = 0$ and all $b_j = 0$, so we assume that some $a_i \neq 0$ or $b_j \neq 0$.

If $n = 1$ then $a_1^2 = b_1^2$ implies $a_1 = b_1$ or $a_1 = -b_1$, but since $a_1^3 = b_1^3$, we have $a_1 = b_1$.

The result will now follow by induction on n if we show that the assumption of $\sum_{i=1}^{n} a_i^r = \sum_{i=1}^{n} b_i^r$ holding for integers $r \geq 2$ implies that $a_i = b_j$ for some i and j.

Suppose (relabeling as needed) that we have distinct values a_i, $1 \leq i \leq s$, with a_i occuring m_i times, and similarly distinct values b_i, $1 \leq i \leq t$, with b_i appearing n_i times. Thus we have

$$\sum_{i=1}^{s} m_i a_i^r = \sum_{i=1}^{t} n_i b_i^r$$

for $r \geq 2$.

Consider the $s + t$ by $s + t$ homogeneous linear system whose $(r+1)$st equation is

$$\sum_{i=1}^{s} a_i^r x_i + \sum_{i=1}^{t} b_i^r x_{s+i} = 0,$$

$0 \leq r \leq s + t - 1$. The system has a nontrivial solution, namely, $x_i = m_i a_i^2$ for $1 \leq i \leq s$ and $x_{s+i} = -n_i b_i^2$ for $1 \leq i \leq t$. Thus the (Vandermonde) determinant of its coefficient matrix is zero, i.e.,

$$\prod_{1 \leq i < j \leq s} (a_j - a_i) \cdot \prod_{1 \leq i < j \leq t} (b_j - b_i) \cdot \prod_{\substack{1 \leq i \leq s \\ 1 \leq j \leq t}} (b_j - a_i) = 0.$$

Since the a_i are distinct and the b_i are distinct, we must have $a_i = b_j$ for some i and j.

20.19 First we obtain the result

$$\sum_{j=1}^{n} \frac{1}{a_j} = a_n + a_{n+1} - 1$$

which is straightforward to prove by induction or can be "discovered" as follows:

Let

$$f(x) = \sum_{j=1}^{n} \frac{1}{a_j} x^j.$$

Then

$$f'(x) = \sum_{j=1}^{n} \frac{j}{a_j} x^{j-1}$$

and so

$$x f'(x) = \sum_{j=1}^{n} \frac{j}{a_j} x^j.$$

Integrating (by parts) gives

$$
\begin{aligned}
x f(x) - \int f(x)\, dx &= \sum_{j=1}^{n} \frac{j}{j+1} \cdot \frac{1}{a_j} x^{j+1} + C \\
&= \sum_{j=1}^{n} \frac{1}{a_{j+2}} x^{j+1} + C \\
&= \frac{1}{x} \sum_{j=1}^{n} \frac{1}{a_{j+2}} x^{j+2} + C \\
&= \frac{1}{x} \left[f(x) - x - x^2 + \frac{1}{a_{n+1}} x^{n+1} + \frac{1}{a_{n+2}} x^{n+2} \right] \\
&\quad + C
\end{aligned}
$$

and so

$$\int f(x)\, dx = \left(x - \frac{1}{x} \right) f(x) + 1 + x - \frac{1}{a_{n+1}} x^n - \frac{1}{a_{n+2}} x^{n+1} - C$$

and thus

$$f(x) = \left(x - \frac{1}{x} \right) f'(x) + \left(1 + \frac{1}{x^2} \right) f(x) + 1 - \frac{n}{a_{n+1}} x^{n-1} - \frac{n+1}{a_{n+2}} x^n.$$

Evaluating at $x = 1$ gives

$$\sum_{j=1}^{n} \frac{1}{a_j} = f(1) = \frac{n}{a_{n+1}} + \frac{n+1}{a_{n+2}} - 1 = a_n + a_{n+1} - 1.$$

Now, by iteration,

$$a_{2n+1} = \frac{2}{1} \cdot \frac{4}{3} \cdot \cdots \cdot \frac{2n}{2n-1}$$

and thus by Wallis' formula (see Glossary) we see that

$$\frac{a_{2n+1}}{\sqrt{2n}} \longrightarrow \sqrt{\frac{\pi}{2}}.$$

and

$$\frac{a_{2n}}{\sqrt{2n}} = \frac{\sqrt{2n}}{a_{2n+1}} \longrightarrow \sqrt{\frac{2}{\pi}}.$$

Therefore, if

$$x_n = \frac{1}{\sqrt{n}} \sum_{j=1}^{n} \frac{1}{a_j},$$

then

$$x_{2n} = \frac{1}{\sqrt{2n}} (a_{2n} + a_{2n+1} - 1) \longrightarrow \sqrt{\frac{\pi}{2}} + \sqrt{\frac{2}{\pi}}.$$

Similarly, we also find that

$$x_{2n+1} \longrightarrow \sqrt{\frac{\pi}{2}} + \sqrt{\frac{2}{\pi}},$$

and hence

$$x_n \longrightarrow \sqrt{\frac{\pi}{2}} + \sqrt{\frac{2}{\pi}}.$$

Additional Problems

20.20 Show that the square root of $0.0\overline{123456790}$ is $0.\overline{1}$.

20.21 Two students, Alpha and Beta, are taking different classes but will each take four tests. They decide to compete to see who can earn the higher total percentage score. Alpha obtains a higher percentage score than Beta on each of four tests. Does it follow that Alpha has a higher total percentage score than Beta?

20.22 (*The Pentagon*, Problem 414, Fall 1988) Let S denote the sum of the infinite series $1 + 1/2 + 1/4 + 1/5 + 1/8 + 1/10 + \cdots$, in which each denominator contains no prime factor except 2 or 5. What is the value of S?

20.23 With a two-pan balance, how many weights are needed to weigh objects of every integral weight from 1 to n kilograms?

20.24 Give an example of an increasing function $f: \mathbf{R} \to \mathbf{R}$ which is discontinuous precisely at the rational numbers.

20.25 Find
$$\max_{x>0} \left(\sin x + \sin \frac{1}{x} \right).$$

20.26 Let S be a semigroup (i.e., a set with an associative binary operation) such that for each $x \in S$ there exists a unique $y \in S$ with $xyx = x$. Show that S is a group.

20.27 Let I, R be subsets of $\mathbf{C}$ closed under addition with I contained in R. Suppose also that $a \in I$ and $x \in R$ imply $ax \in I$. Prove that the set $\{x \in R : x^n \in I \text{ for some } n \geq 1\}$ is closed under addition.

20.28 Show that the power of 2 in $n!$ is n minus the sum of the binary digits in n.

Note: This result is a special case of the following theorem of Legendre. (See [18, vol. 1, pages 101–102].) Let n be a positive integer and p a prime number. Then the power of p in $n!$ is

$$\frac{n-s}{p-1}$$

where s is the sum of the digits in the base-p representation of n.

20.29 Show that
$$\frac{[m!e] - [n!e]}{m-n}$$
is an integer for all positive integers m, n with $m \neq n$.

20.30 Find the two four-digit numbers x such that x^k has the same four last digits for every $k \geq 1$.

20.31 Let L be a line in the x-y plane having rational slope. Show that there exists a positive real number d such that any point in the plane with integer coordinates, not on L, is at distance at least d from L.

20.32 Given four distinct integers a_1, a_2, a_3, a_4, show that there is a set S of density $1/3$ such that every integer is of the form $s + a_i$, $s \in S$.

Note: A set S has *density d* if
$$\lim_{n \to \infty} \frac{|S \cap \{1, \ldots, n\}|}{n} = d.$$

20.33 Let a, b, c be integers such that the line $ax + by = c$ passes through an integer lattice point. Show that it passes through one with

$$\max(|x|, |y|) \leq \max(|a|, |b|, |c|).$$

20.34 Prove that there are infinitely many positive integers n that cannot be written in the form $x^2 + y^3 + z^6$ for nonnegative integers x, y, z.

20.35 Suppose that a, b, c are real, $a > 0$, $c > 0$, and $ac > b^2$. Show that

$$\frac{ax^2 + 2bxy + cy^2}{x^2 + y^2}$$

has positive upper and lower bounds for all ordered pairs (x, y) of real numbers not both 0. What are the least upper bound and greatest lower bound?

20.36 Let $F_n = 2^{2^n} + 1$. Prove that no two of the numbers F_0, F_1, ... have a prime factor in common.

Note: The numbers F_n are called *Fermat numbers*.

20.37 Let $E_0 = 2$, $E_1 = 2 + 1 = 3$, $E_2 = 2 \cdot 3 + 1 = 7$, $E_3 = 2 \cdot 3 \cdot 7 + 1 = 43$, $E_4 = 2 \cdot 3 \cdot 7 \cdot 43 + 1 = 1807$, $E_5 = 2 \cdot 3 \cdot 7 \cdot 43 \cdot 1807 + 1$, ... Prove that no two of the numbers E_0, E_1, E_2, E_3, ... have a prime factor in common.

20.38 (a) If p is a prime number and n is any positive integer, show that

$$\left(1 - \frac{1}{p}\right)^{-1} > 1 + \frac{1}{p} + \frac{1}{p^2} + \cdots + \frac{1}{p^n}.$$

(b) If N is a positive integer greater than 1 and p_1, p_2, ..., p_k are distinct primes not exceeding N, prove that

$$\left(1 - \frac{1}{p_1}\right)^{-1} \left(1 - \frac{1}{p_2}\right)^{-1} \cdots \left(1 - \frac{1}{p_k}\right)^{-1} > 1 + \frac{1}{2} + \frac{1}{3} + \cdots + \frac{1}{N}.$$

Hint: Use (a) for a suitable value of n.

(c) Use (b) to demonstrate that there are infinitely many primes.

20.39 Prove that the sequence of integers 1, 5, 13, 29, 61, ..., where the k-th term is $2^{k+1} - 3$, contains an infinite subsequence no two terms of which have a prime factor in common.

20.40 Find all $n \in \mathbf{N}$ such that all prime factors of $9^n + 1$ are less than 40.

20.41 (Putnam Competition, 1978) Let n distinct points in the plane be given. Prove that fewer than $2n^{3/2}$ pairs of them are unit distance apart.

20.42 Use calculus to find the maximum of $xy(1 - x - y)$ for (x, y) in the triangular region defined by the inequalities $x \geq 0$, $y \geq 0$, and $x + y \leq 1$.

20.43 (a) Show that a convex curve in the plane of arc length S can have at most $S + 1$ lattice points on it.

(b) Show that a strictly convex curve in the plane of arc length S can have at most $cS^{2/3}$ lattice points on it, where c is a constant.

20.44 For odd $n \geq 1$, find all real θ such that

$$\sum_{k=0}^{(n-1)/2} (-1)^n \binom{n}{2k} (\tan \theta)^{n-1-2k} = 0.$$

20.45 Show that for any integer m, and any number $n > 1$,

$$\sum_{k=1}^{n-1} k \cos \frac{2\pi km}{n} = -\frac{n}{2}$$

and

$$\sum_{k=1}^{n-1} k \sin \frac{2\pi km}{n} = -\frac{n}{2} \cot \frac{\pi m}{n}.$$

20.46 Show that

$$\sum_{n=0}^{\infty} \frac{(-1)^n \pi^{6n}}{(6n)!} = -\frac{1}{3}$$

and

$$\sum_{n=0}^{\infty} \frac{(-1)^n \pi^{4n}}{2^{2n}(4n)!} = 0.$$

20.47 (Bulgarian Olympiad) In a plane there are n circles each of unit radius. Prove that at least one of these circles contains an arc that does not intersect any of the other circles and whose length is not less than $2\pi/n$.

20.48 Show that if $n^n + 1$ is an odd prime, then $n = 2^{2^k}$ for some nonnegative integer k.

20.49 Suppose that a, b, and c are distinct integers and P is a polynomial with integral coefficients. Prove that it is impossible to have $a = P(b)$, $b = P(c)$, and $c = P(a)$.

20.50 What is the rightmost nonzero digit in $1000000!$?

20.51 (Putnam Competition, 1982) Assume that the system of simultaneous differential equations

$$y' = -z^3, \ z' = y^3$$

with the initial conditions $y(0) = 1$, $z(0) = 0$ has a unique solution $y = f(x)$, $z = g(x)$ defined for all real x. Prove that there exists a positive constant L such that for all real x, $f(x+L) = f(x)$, $g(x+L) = g(x)$.

20.52 Find conditions on the parameters a, b, c, d so that

$$f(x,y) = a\sin(x+y) + b\cos(x+y) + c\sin(x-y) + d\cos(x-y)$$

can be written as $f(x,y) = g(x)h(y)$.

20.53 A point P is in the interior of a circle of radius r. Place the vertex of a right angle at P and denote by A and B the points where the sides of the right angle intersect the circle. Let Q be the point which completes the rectangle $PAQB$. What is the locus of Q?

20.54 Find positive integers n and $a_1, a_2, \ldots, a_n$ such that

$$a_1 + a_2 + \cdots + a_n = 1999$$

and the product $a_1 a_2 \ldots a_n$ is as large as possible.

20.55 (*Math. Magazine*, Problem 1308, December 1988) Let $\mathbf{N}$ be the set of natural numbers $\{1, 2, 3, \ldots\}$, let $g : \mathbf{N} \longrightarrow \mathbf{N}$ be a bijection, and $a \in \mathbf{N}$ be an odd number.

(a) Prove that there is no function f such that $f(f(n)) = g(n) + a$ for all $n \in \mathbf{N}$.

(b) What if a is even?

20.56 (*American Math. Monthly*, Problem E3310, February 1989) I have a secret positive integer u, not exceeding some predetermined bound N. If you give me any positive integer n, I will tell you whether or not $u + n$ is prime.

(a) Give a number $G(N)$ and a procedure such that you can always determine the value of u in at most $G(N)$ such trials.

(b) Give a number $F(N)$ such that no strategy will always determine u in fewer than $F(N)$ such trials.

20.57 (*College Math. Journal*, Problem 391, January 1989) Determine all integral solutions of the equation

$$\frac{2x}{1-x^2} + \frac{2y}{1-y^2} + \frac{2z}{1-z^2} = \frac{8xyz}{(1-x^2)(1-y^2)(1-z^2)}.$$

20.58 Find a function $g(x)$, continuous on $(0,1]$ with $g(1) = 0$, such that

$$h(x) = g(x) + \sum_{n=0}^{\infty} x^{2^n}$$

satisfies $h(x^2) = h(x) - 1$ for all x in $(0,1)$.

20.59 An infinite set S of points in the plane has the property that the distance between any two of them is an integer. Prove that the points of S lie on a single straight line.

20.60 Let I_n be the n by n multiplicative identity matrix and let J_n be the n by n matrix consisting of all 1's. Evaluate the determinant of the matrix $1989 \cdot I_{1988} - J_{1988}$.

20.61 Define functions $\phi_n(x)$ by $\phi_0(x) = x$,

$$\phi_{n+1}(x) = (\phi_n(x))^2 - 2$$

for $n \geq 0$. Let $x \in [-2, 2]$ be such that $\phi_n(x) \neq 0$ for all $n \geq 0$, and suppose that $\epsilon_n = \pm 1$ has the same sign as $\phi_n(x)$. Prove that

$$\epsilon_0 \sqrt{2 + \epsilon_1 \sqrt{2 + \cdots + \epsilon_{n-1}\sqrt{2}}} \to x$$

as $n \to \infty$.

20.62 Let $\psi(x) = \frac{1}{4}x(x^3 + 16)/(x^3 - 2)$. Suppose that x is a rational number and $x^3 - 2$ is the square of a rational number.

(a) Show that $(\psi(x))^3 - 2$ is also the square of a rational number.

(b) Show that the numbers 3, $\psi(3)$, $\psi(\psi(3))$, ... are all different.

Note: These results show that there are infinitely many rational numbers x, y with $y^2 = x^3 - 2$.

20.63 Find all square matrices A such that all elements of both A and of A^{-1} are nonnegative. Prove that there are no others.

Glossary

Abel's formula. The relation

$$\sum_{k=1}^{n} a_k b_k = A_n b_n - \sum_{k=1}^{n-1} A_k (b_{k+1} - b_k),$$

for $n \geq 2$, where $A_k = \sum_{i=1}^{k} a_i$.

Absolutely convergent series. A series $\sum a_n$ for which $\sum |a_n|$ converges.

AM, GM, HM, and AM–GM–HM inequality. Let $x_1, \ldots, x_n$ and $w_1, \ldots, w_n$ be nonnegative real numbers, and let $w = \sum_{i=1}^{n} w_i$. The *arithmetic mean* (AM), *geometric mean* (GM), and *harmonic mean* (HM) of $x_1, \ldots, x_n$ with *weights* $w_1, \ldots, w_n$ are defined as

$$\mathrm{AM} = \frac{1}{w} \sum_{i=1}^{n} w_i a_i$$

$$\mathrm{GM} = \left(\prod_{i=1}^{n} a_i^{w_i} \right)^{1/w}$$

$$\mathrm{HM} = \frac{w}{\sum_{i=1}^{n} w_i / a_i}.$$

These means satisfy the inequalities

$$\mathrm{HM} \leq \mathrm{GM} \leq \mathrm{AM}.$$

Bertrand's postulate. For every $n \geq 2$, there exists a prime number between n and $2n$.

Binomial coefficient. The expression

$$\binom{n}{k} = \frac{n!}{k!(n-k)!},$$

237

for $0 \le k \le n$. The value of this expression is the number of k-element subsets of an n-element set. Another important binomial coefficient expression is

$$\binom{n+k-1}{k-1},$$

which equals the number of distributions of n indistinguishable objects into k classes.

Binomial theorem. For $n \ge 0$,

$$(x+y)^n = \sum_{k=0}^{n} \binom{n}{k} x^k y^{n-k}.$$

Letting $x = y = 1$, we obtain the identity

$$2^n = \sum_{k=0}^{n} \binom{n}{k}.$$

Cauchy's criterion. A real-valued sequence $\{s_n\}$ converges if, for each $\epsilon > 0$, there exists an $N \ge 1$ such that $|s_m - s_n| < \epsilon$ for all $m \ge N$ and $n \ge N$.

Cauchy's formula. Suppose that $f(z)$ is analytic in an open set D and C is a simple, closed contour in the interior of D. Given that the Taylor series expansion of $f(z)$ is $\sum a_n z^n$, then, for any point z_0 in the interior of C, and $n \ge 0$,

$$a_n = \frac{1}{2\pi i} \int_C \frac{f(z)}{(z-z_0)^{n+1}} \, dz.$$

Convex hull. The smallest convex region that contains a given set of points in the plane or in space. For a finite set S of points in the plane, the convex hull is a polygon whose vertices are members of S, or else a line segment whose endpoints are members of S.

Countable and uncountable sets. A set S is *countable* if it is finite or it can be put into one-to-one correspondence with the set $\mathbf{N}$ of natural numbers. Otherwise, S is *uncountable*. Examples: the sets $\mathbf{N}$, $\mathbf{Z}$, and $\mathbf{Q}$ are countable, while the sets $\mathbf{R}$ and $\mathbf{C}$ are uncountable.

De Moivre's theorem. Let $z = r(\cos\theta + i\sin\theta)$. Then, for $n \ge 0$,

$$z^n = r^n(\cos n\theta + i\sin n\theta).$$

Digraph (directed graph). A graph in which each edge $\{x, y\}$ has been given a direction, i.e., from x to y or from y to x.

Euler's formula. Suppose that a connected planar graph with V vertices and E edges partitions the plane into F regions ("faces"). Then

$$V + F = E + 2.$$

Note: Euler's formula is easily deduced from Pick's theorem, and *vice versa*.

Euler's ϕ function. For $n \geq 1$, $\phi(n)$ is the number of integers between 1 and n that are relatively prime to n. If the factorization of n into primes is

$$n = \prod_{i=1}^{k} p_i^{\alpha_i},$$

then

$$\phi(n) = \prod_{i=1}^{k} \left(p_i^{\alpha_i} - p_i^{\alpha_i - 1} \right).$$

The following fact is useful:

$$\sum_{d \mid n} \phi(d) = n.$$

Euler's theorem. If $\gcd(a, m) = 1$, then

$$a^{\phi(m)} \equiv 1 \bmod m.$$

Note: If the modulus m is prime, we obtain Fermat's (little) theorem.

Fermat's (little) theorem. If p is a prime number and $p \nmid a$, then

$$a^{p-1} \equiv 1 \bmod p.$$

Field. A *field* F is a set (with at least two elements) on which are defined two binary operations, $+$ and $\cdot$, such that the following conditions hold:

1. F is an abelian group with respect to $+$;
2. $F - \{0\}$ (where 0 is the additive identity) is an abelian group with respect to $\cdot$;

3. (distributive law) for all x, y, $z \in F$,

$$x \cdot (y + z) = x \cdot y + x \cdot z.$$

Examples: (1) the set of $\mathbf{R}$ of real numbers with the usual addition and multiplication; (2) the set $\mathbf{Q}$ of rational numbers; (3) the set $\mathbf{Z}_2 = \{0, 1\}$ with addition and multiplication modulo 2.

Fibonacci sequence. The sequence

$$\{f_n\} = \{0, 1, 1, 2, 3, 5, 8, 13, 21, 34, \ldots\}$$

defined by $f_0 = 0$, $f_1 = 1$ and $f_n = f_{n-1} + f_{n-2}$ for $n \geq 2$.

Graph. A *graph* G consists of a set V of *vertices* and a set E of *edges* joining some pairs of vertices. Vertices joined by an edge are said to *adjacent*. Vertices not joined by an edge are *nonadjacent*. In a drawing of a graph, adjacent vertices may be joined by a straight or curved line and the lines may cross arbitrarily.

The picture below shows a graph with seven vertices and ten edges.

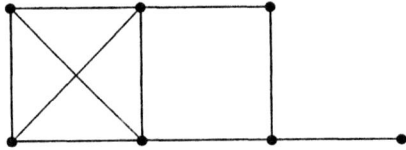

The *degree* of a vertex v is the number of vertices adjacent to v. In the above graph, the vertices have degrees 1, 2, 3, 3, 3, 4, and 4.

Greatest common divisor. Let a and b be integers, not both 0. The *greatest common divisor* of a and b, written $\gcd(a, b)$, is the greatest integer that divides both a and b. Given integers a and b with greatest common divisor g, there exist integers x and y such that

$$ax + by = g.$$

Group. A *group* G is a nonempty set on which is defined a binary operation $*$ satisfying the following three laws:

1. (associativity) for all x, y, $z \in G$, $x * (y * z) = (x * y) * z$;
2. (identity element) G contains an element e with the property that, for all $x \in G$, $x * e = e * x = x$;

3. (inverse elements) for every $x \in G$ there exists an $x^{-1} \in G$ with the property that $x * x^{-1} = x^{-1} * x = e$.

Examples: (1) the set of integers $\mathbf{Z}$ under addition; (2) the set of nonzero real numbers $\mathbf{R} - \{0\}$ under multiplication; (3) the set of invertible $n \times n$ matrices with real entries under matrix multiplication.

We say that G is *abelian* if $x * y = y * x$ for all $x, y \in G$; otherwise, G is *nonabelian*. Examples: the groups (1) and (2) above are abelian and (3) is nonabelian.

Intermediate value theorem. If f is a continuous real-valued function defined on the closed interval $[a, b]$, and y is any number between $f(a)$ and $f(b)$, then there exists a number x in (a, b) such that $f(x) = y$.

Jordan curve theorem. If C is a simple closed curve in the plane, then C divides the plane into two components.

Lattice point. An ordered n-tuple $(x_1, \ldots, x_n) \in \mathbf{R}^n$ where $x_1, \ldots, x_n$ are integers.

Law of cosines. Given $\triangle ABC$ labeled as below,

$$c^2 = a^2 + b^2 - 2ab \cos C.$$

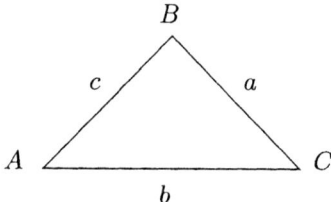

Law of sines. Given $\triangle ABC$ labeled as above,

$$\frac{\sin A}{a} = \frac{\sin B}{b} = \frac{\sin C}{c}.$$

Lebesgue's dominated convergence theorem. If $f_n \to f$ for $x \in (a, b)$, and $f_n(x) \le g(x)$ for all $x \in (a, b)$ and $n \ge 0$, then

$$\lim_{n \to \infty} \int_a^b f_n(x)\, dx = \int_a^b f(x)\, dx.$$

L'Hôpital's rule. Suppose that f and g are differentiable functions and g' is nonzero on an interval containing a (except possibly at a). If $\lim_{x\to a} f(x) = 0$ and $\lim_{x\to a} g(x) = 0$, or if $\lim_{x\to a} f(x) = \pm\infty$ and $\lim_{x\to a} g(x) = \pm\infty$, then

$$\lim_{x\to a} \frac{f(x)}{g(x)} = \lim_{x\to a} \frac{f'(x)}{g'(x)}$$

(if the second limit exists).

Numbers. The set of natural numbers $\{1, 2, 3, \ldots\}$ is denoted by $\mathbf{N}$. The set of integers $\{\ldots, -3, -2, -1, 0, 1, 2, 3, \ldots\}$ is denoted by $\mathbf{Z}$.

The set of real numbers is denoted by $\mathbf{R}$. A real number is *rational* if it is of the form p/q, where p and q are integers and $q \neq 0$. The set of rational numbers is denoted by $\mathbf{Q}$. A real number that is not rational is called *irrational*. Examples: $2/3$, $-4/7$, and 14 are rational numbers, while $\sqrt{2}$, e, and π are irrational.

The set of complex numbers is denoted by $\mathbf{C}$.

A complex number is *algebraic* if it is the root of a polynomial with integer coefficients. A number that is not algebraic is called *transcendental*. Examples: $\sqrt{2}$ is algebraic and π and e are transcendental.

Pascal's identity and triangle. For $1 \leq k \leq n$,

$$\binom{n+1}{k} = \binom{n}{k} + \binom{n}{k-1}.$$

Pascal's identity is the basis for generating Pascal's triangle of binomial coefficients:

$$
\begin{array}{ccccccccccc}
 & & & & & 1 & & & & & \\
 & & & & 1 & & 1 & & & & \\
 & & & 1 & & 2 & & 1 & & & \\
 & & 1 & & 3 & & 3 & & 1 & & \\
 & 1 & & 4 & & 6 & & 4 & & 1 & \\
1 & & 5 & & 10 & & 10 & & 5 & & 1 \\
 & & & & & \vdots & & & & &
\end{array}
$$

Perfect number. A number n the sum of whose divisors equals $2n$. In general, the sum of the divisors of n is denoted $\sigma(n)$. Thus a number is perfect if $\sigma(n) = 2n$. We say that n is *abundant* if $\sigma(n) > 2n$ and *deficient* if $\sigma(n) < 2n$. Examples: 6 and 28 are perfect numbers; 10 and 15 are deficient numbers; and 36 and 100 are abundant numbers.

Note: If the factorization of n into primes is

$$n = \prod_{i=1}^{k} p_i^{\alpha_i},$$

then

$$\sigma(n) = \prod_{i=1}^{k} \left(\frac{p_i^{\alpha_i+1} - 1}{p_i - 1} \right).$$

Pick's theorem. A non-self-intersecting polygon with lattice point vertices has area $A = b/2 + i - 1$, where b and i are the number of lattice points on the boundary and in the interior of the polygon, respectively.

Prime and composite numbers. A natural number $n \geq 2$ is *composite* if n has positive factors other than 1 and n. Otherwise, n is *prime*. Examples: 15, 100, and 2^{100} are composite numbers; 2, 17, and 1999 are primes.

Prime number theorem. Let $\pi(n)$ be the number of primes less than or equal to n. Then

$$\pi(n) \sim n/\ln n.$$

Stirling's approximation:

$$n! \sim n^n e^{-n} \sqrt{2\pi n}.$$

Triangle. A triangle has four basic "centers." (1) The *circumcenter* is the center of the triangle's circumscribed circle. It is the intersection of the perpendicular bisectors of the sides of the triangle. (2) The *incenter* is the center of the triangle's inscribed circle. It is the intersection of the three angle bisectors of the triangle. (3) The *centroid* is the center of mass of the triangle. It is the intersection of the three medians of the triangle (a *median* is a line joining a vertex to the midpoint of the opposite side). (4) The *orthocenter* is the intersection of the three altitudes of the triangle (an *altitude* is a line passing through a vertex and perpendicular to the opposite side).

The area of a triangle is $A = \frac{1}{2}ah$, where a is a side length and h is the length of the altitude from the opposite vertex. The area is also given by *Heron's formula*:

$$A = \sqrt{s(s-a)(s-b)(s-c)},$$

where a, b, c are the side lengths and $s = \frac{1}{2}(a+b+c)$ (s is called the *semiperimeter*).

In any triangle with side lengths a, b, and c, the *triangle inequality* asserts that

$$a < b + c.$$

If the sides are given by vectors x, y, and $x+y$, the inequality becomes

$$|x + y| \leq |x| + |y|.$$

Equality occurs when the vectors are parallel (and the triangle degenerate).

Tournament. A complete digraph, i.e., one in which every pair of vertices x, y are joined by an arrow from x to y or from y to x.

Uniform continuity. A real-valued function f is *uniformly continuous* on a subset $A \subseteq \mathbf{R}$ if, for each $\epsilon > 0$, there exists a $\delta > 0$ such that $|f(x) - f(y)| < \epsilon$ for all x, $y \in A$ for which $|x - y| < \delta$.

Note: A real-valued function f is *continuous* on a subset $A \subseteq \mathbf{R}$ if, for each $x \in A$ and each $\epsilon > 0$ there exists a $\delta > 0$ such that $|f(x) - f(y)| < \epsilon$ for all $y \in A$ for which $|x - y| < \delta$. Since the value of δ depends on both x and ϵ, point-wise continuity is a weaker condition than uniform continuity.

Uniform Convergence. A sequence of functions $\{f_n\}$ *converges uniformly* to a function f on the set A provided that for every $\epsilon > 0$, there exists $n_0 \geq 1$ such that $|f_n(x) - f(x)| < \epsilon$ for all $n \geq n_0$ and all $x \in A$.

A series $\sum_{n=1}^{\infty} f_n(x)$ converges uniformly to a function $s(x)$ on A provided that the sequence $\{s_n\}$ converges uniformly to s on A, where $s_n = \sum_{i=1}^{n} f_i$.

Vandermonde determinant formula. For $n \geq 1$,

$$V_n = \prod_{1 \leq i < j \leq n} (r_j - r_i),$$

where

$$V_n = \begin{vmatrix} 1 & 1 & \cdots & 1 \\ r_1 & r_2 & \cdots & r_n \\ r_1^2 & r_2^2 & \cdots & r_n^2 \\ \vdots & \vdots & \vdots & \vdots \\ r_1^{n-1} & r_2^{n-1} & \cdots & r_n^{n-1} \end{vmatrix}.$$

Vandermonde's identity. For $0 \leq m \leq n$,

$$\binom{m+n}{m} = \sum_{k=0}^{m} \binom{m}{k}\binom{n}{m-k}.$$

Vector space. A *vector space* V over a field F is an additive abelian group together with a rule which assigns to every $f \in F$ and $v \in V$ an element $f \cdot v \in V$ such that, for all f, f_1, $f_2 \in F$ and v, v_1, $v_2 \in V$, the following conditions hold:

1. $f \cdot (v_1 + v_2) = f \cdot v_1 + f \cdot v_2$;
2. $(f_1 + f_2) \cdot v = f_1 \cdot v + f_2 \cdot v$;
3. $f_1 \cdot (f_2 \cdot v) = (f_1 f_2) \cdot v$;
4. $1 \cdot v = v$, where 1 is the multiplicative identity of F.

The elements of V are called *vectors* and the elements of F are called *scalars*.

Examples: (1) the group $\mathbf{R}^2$ is a vector space over the field $\mathbf{R}$; (2) the group $\mathbf{R}$ is a vector space over the field $\mathbf{Q}$ of rational numbers; (3) the group of continuous functions from $\mathbf{R}$ to $\mathbf{R}$ is a vector space over $\mathbf{R}$.

Wallis' formula. As $n \to \infty$,

$$\frac{2}{1} \cdot \frac{4}{3} \cdot \ \cdots \ \cdot \frac{2n}{2n-1} \cdot \frac{1}{\sqrt{2n+1}} \ \longrightarrow \ \sqrt{\frac{\pi}{2}}.$$

Weierstrass M-test. Suppose that $\{f_n\}$ is a sequence of real-valued functions defined on an interval I, and M_n are constants such that $\sum M_n$ converges and $|f_n(x)| \leq M_n$ for all $x \in I$ and $n \geq 1$. Then $\sum f_n$ converges uniformly on I.

Wilson's theorem. If p is prime, then

$$(p-1)! \equiv -1 \ \mathrm{mod} \ p.$$

Bibliography

[1] G. L. Alexanderson, L. F. Klosinski, and L. C. Larson. *The William Lowell Putnam Mathematical Competition, Problems and Solutions 1965-1984*. Mathematical Association of America, Washington, D. C., 1985.

[2] E. Barbeau, M. Klamkin, and W. Moser. *Five Hundred Mathematical Challenges*. Mathematical Association of America, Washington, D. C., 1995.

[3] D. Cox, J. Little, and D. O'Shea. *Ideals, Varieties, and Algorithms*. Springer–Verlag, New York, 1992.

[4] E. R. Emmet. *Puzzles for Pleasure*. Barnes & Noble Books, New York, 1995.

[5] A. M. Gleason, R. E. Greenwood, and L. M. Kelly. *The William Lowell Putnam Mathematical Competition, Problems and Solutions 1938-1964*. Mathematical Association of America, Washington, D. C., 1980.

[6] S. L. Greitzer. *International Mathematical Olympiads 1959-1977*. Mathematical Association of America, Washington, D. C., 1978.

[7] M. Hall. *Combinatorial Theory*. Wiley, New York, second edition, 1986.

[8] K. Hardy and K. Williams. *The Green Book of Mathematical Problems*. Dover, New York, 1985.

[9] I. N. Herstein. *Abstract Algebra*. Prentice–Hall, Upper Saddle River, third edition, 1996.

[10] R. Honsberger. *In Pólya's Footsteps*. Mathematical Association of America, Washington, D. C., 1997.

[11] R. Johnsonbaugh. *Discrete Mathematics*. Macmillan Publishing Company, New York, third edition, 1993.

[12] Murray S. Klamkin. *International Mathematical Olympiads 1978–1985 and Forty Supplemental Problems*. Mathematical Association of America, Washington, D. C., 1986.

[13] V. Klee and S. Wagon. *Old and New Unsolved Problems in Plane Geometry and Number Theory*. Mathematical Association of America, Ithaca, 1991.

[14] L. Larson. *Problem-Solving Through Problems*. Springer-Verlag, New York, 1983.

[15] E. Lozansky and C. Rousseau. *Winning Solutions*. Springer-Verlag, New York, 1996.

[16] D. J. Newman. *A Problem Seminar*. Springer–Verlag, New York, 1982.

[17] G. Pólya. *How to Solve It*. Doubleday, New York, 1957.

[18] E. Rapaport. *Hungarian Problem Book*, volume 1 and 2. Random House, New York, 1963.

[19] W. Rudin. *Principles of Mathematical Analysis*. McGraw–Hill, Inc., New York, 1976.

[20] J. Székely. *Contests in Higher Mathematics*. Springer–Verlag, New York, 1996.

Index